Johannes Karl Fischer

Scaling Laws of Optical Fibre Nonlinearity

Johannes Karl Fischer

Scaling Laws of Optical Fibre Nonlinearity

in Dispersion-Managed Transmission Systems

Südwestdeutscher Verlag für Hochschulschriften

Impressum/Imprint (nur für Deutschland/ only for Germany)
Bibliografische Information der Deutschen Nationalbibliothek: Die Deutsche Nationalbibliothek verzeichnet diese Publikation in der Deutschen Nationalbibliografie; detaillierte bibliografische Daten sind im Internet über http://dnb.d-nb.de abrufbar.

Alle in diesem Buch genannten Marken und Produktnamen unterliegen warenzeichen-, marken- oder patentrechtlichem Schutz bzw. sind Warenzeichen oder eingetragene Warenzeichen der jeweiligen Inhaber. Die Wiedergabe von Marken, Produktnamen, Gebrauchsnamen, Handelsnamen, Warenbezeichnungen u.s.w. in diesem Werk berechtigt auch ohne besondere Kennzeichnung nicht zu der Annahme, dass solche Namen im Sinne der Warenzeichen- und Markenschutzgesetzgebung als frei zu betrachten wären und daher von jedermann benutzt werden dürften.

Verlag: Südwestdeutscher Verlag für Hochschulschriften Aktiengesellschaft & Co. KG
Dudweiler Landstr. 99, 66123 Saarbrücken, Deutschland
Telefon +49 681 37 20 271-1, Telefax +49 681 37 20 271-0
Email: info@svh-verlag.de
Zugl.: Berlin, Technische Universität Berlin, Dissertation, 2009

Herstellung in Deutschland:
Schaltungsdienst Lange o.H.G., Berlin
Books on Demand GmbH, Norderstedt
Reha GmbH, Saarbrücken
Amazon Distribution GmbH, Leipzig
ISBN: 978-3-8381-1402-6

Imprint (only for USA, GB)
Bibliographic information published by the Deutsche Nationalbibliothek: The Deutsche Nationalbibliothek lists this publication in the Deutsche Nationalbibliografie; detailed bibliographic data are available in the Internet at http://dnb.d-nb.de.

Any brand names and product names mentioned in this book are subject to trademark, brand or patent protection and are trademarks or registered trademarks of their respective holders. The use of brand names, product names, common names, trade names, product descriptions etc. even without a particular marking in this works is in no way to be construed to mean that such names may be regarded as unrestricted in respect of trademark and brand protection legislation and could thus be used by anyone.

Publisher: Südwestdeutscher Verlag für Hochschulschriften Aktiengesellschaft & Co. KG
Dudweiler Landstr. 99, 66123 Saarbrücken, Germany
Phone +49 681 37 20 271-1, Fax +49 681 37 20 271-0
Email: info@svh-verlag.de

Printed in the U.S.A.
Printed in the U.K. by (see last page)
ISBN: 978-3-8381-1402-6

Copyright © 2010 by the author and Südwestdeutscher Verlag für Hochschulschriften Aktiengesellschaft & Co. KG and licensors
All rights reserved. Saarbrücken 2010

Danksagung

Ich möchte mich an dieser Stelle bei all jenen Menschen bedanken, die durch ihre kontinuierliche Unterstützung in den vergangenen fünf Jahren zum Gelingen dieser Arbeit beigetragen haben.

Ganz besonders möchte ich mich bei Herrn Prof. Dr.-Ing. Petermann für die Betreuung der Promotion bedanken. Sein Vorbild war immer ein Ansporn das weitläufige Feld der optischen Nachrichtentechnik noch umfassender kennenzulernen. Auch die umfangreiche Unterstützung gerade am Beginn der Arbeit hat mir sehr geholfen. Ich erinnere mich noch gut an stundenlange Diskussionen, die erst nach Überprüfung und Klärung auch des letzten Vorzeichens beendet waren.

Mein Dank gilt auch Herrn Prof. Dr.-Ing. Hanik, der freundlicherweise die Arbeit begutachtet hat. Insbesondere die frühzeitige Fertigstellung des Gutachtens ermöglicht mir nun den termingerechten Antritt meiner neuen Stelle.

In diesem Zusammenhang möchte ich auch noch einmal allen Mitgliedern der Prüfungskommission – Herrn Prof. Dr.-Ing. Grallert, Herrn Prof. Dr.-Ing. Petermann und Herrn Prof. Dr.-Ing. Hanik – danken, dass trotz deren vielfältigen Verpflichtungen ein frühzeitiger Termin für die wissenschaftliche Aussprache gefunden werden konnte.

Danken möchte ich auch der guten Seele unseres Fachgebiets, Frau Hamer. Sie hat maßgeblich dafür gesorgt, dass ich alle bürokratischen Aufgaben zügig erledigen konnte, indem Sie immer mit Rat und Tat zur Seite stand.

Meinen Kollegen – Sebastian Randel, Hadrien Louchet, Fabian Kerbstadt, Miro Malach, Christian-Alexander Bunge, Stefan Warm, Christian Weber, Marcus Winter, Torsten Mitze, Carsten Voigt und Lars Zimmermann – möchte ich für die schöne Zeit danken. Besonders die vielen Diskussionen (nicht immer fachlicher Art) und natürlich die Fahrten zu Konferenzen haben mir mit euch immer viel Spaß gemacht.

Mein besonderer Dank geht dabei an Christian Weber und Marcus Winter für das Korrekturlesen einiger Kapitel der Dissertation.

Meiner Mutter Veronika Fischer möchte ich aus tiefstem Herzen danken für ihre Aufopferung bei der Betreuung meiner kleinen Tochter. Ohne diese Hilfe wäre die Dissertation wahrscheinlich bis zum heutigen Tage nicht fertig geworden.

Bei meiner Tochter Nele Luise und meiner Frau Gina Cajar möchte ich mich für die lange Zeit entschuldigen, in der ich kein richtiger Vater und Partner sein konnte und mich bei ihnen für die Geduld bedanken, die sie mir gerade auch in schwierigen Zeiten entgegengebracht haben.

Johannes Karl Fischer

Berlin, den 16. November 2009

Contents

1 **Introduction** 1
2 **Propagation Effects** 5
  2.1 Linear Effects 5
      2.1.1 Fibre Loss 6
      2.1.2 Group-Velocity Dispersion 9
  2.2 Nonlinear Effects 10
      2.2.1 Intrachannel Nonlinearities 12
      2.2.2 Interchannel Nonlinearities 18

3 **Single-Span Model** 23
  3.1 Volterra Series Expansion 25
  3.2 Nonlinear Transfer Function 26
      3.2.1 Nonlinear Transfer Function of a Single Fibre 27
      3.2.2 Nonlinear Transfer Function of Multiple Fibre Sections . . . 30
      3.2.3 Nonlinear Transfer Function of a Single-Span System . . . 34
      3.2.4 Nonlinear Transfer Function of Systems with Single-Periodic Dispersion Maps 35
      3.2.5 Nonlinear Transfer Function of Systems with Randomly Varying Residual Dispersion per Span 40
  3.3 Scaling Laws 42
      3.3.1 3-dB Bandwidth of the Nonlinear Transfer Function 43
      3.3.2 Equivalent Precompensation 48
      3.3.3 Nonlinear Phase Shift 49

4 **Application to the Design of Fibre-Optic Transmission Systems** 51
  4.1 Simulation Model 51
      4.1.1 Transmitter 52
      4.1.2 Transmission Line 55

	4.1.3	Receiver	56

	4.1.3	Receiver	56
	4.1.4	Evaluation of System Performance	59
4.2	Numerical Verification of the Single-Span Model		61
	4.2.1	Single-Span Systems	62
	4.2.2	Multi-Span Systems with Single-Periodic Dispersion Map .	68
4.3	Implications on the Design of Fibre-Optic Transmission Systems . .		78
	4.3.1	Multilevel Phase-Modulated Signals	78
	4.3.2	Varying Spectral Efficiency	88

5 Conclusion — 95
5.1 Summary of Key Results 95
5.2 Further Work or Shortcomings and Pitfalls 99

Acronyms — 102

List of Symbols — 105

List of Figures — 107

Bibliography — 114

1 Introduction

Throughout human history technological advancements have changed the ways in which we communicate. Today, personal mobile communication devices enable access to huge information resources provided by the internet at almost any place at any time. Demand for integration of television broadcasting, voice and data services over internet protocol (IP) requires increased bandwidth per customer and improved flexible architectures of access and backbone networks [1–3]. Furthermore, future services like high-definition television over IP or even distribution of super-high-definition digital cinema over optical fibres [4] need tremendous amounts of transmission bandwidth. Increasing penetration of homes with broadband fibre to the home (FTTH) connections puts a growing strain on the capacity of backbone networks [5,6]. These developments necessitate upgrades of backbone networks.

The capacity of fibre-optic networks can either be increased by utilising existing legacy fibre plant and upgrading terminal equipment only or by building entirely new fibre lines with modern technology. Obviously, the former solution is more cost-efficient, because the infrastructure of buried fibre lines remains untouched. Pursuing this upgrade path allows to replace legacy transmitters and receivers with modernised equipment.

At the transmitter of fibre-optic communication systems, information is modulated onto the light of lasers emitting at different wavelengths. The light from different laser sources is then transmitted together over a single optical fibre. This technique is called wavelength-division multiplexing (WDM) and allows simple extraction of wavelength channels by using optical bandpass filters. This enables different users to share a single fibre line. Furthermore, WDM increases the total capacity tremendously compared to transmission over a single wavelength channel. The capacity of a WDM system can be increased by adding additional wavelength channels or by increasing the bit rate per WDM channel. However, the total number of wavelength channels is eventually limited by the available bandwidth of the optical inline amplifiers used to periodically amplify the optical signal along the transmission line.

1 Introduction

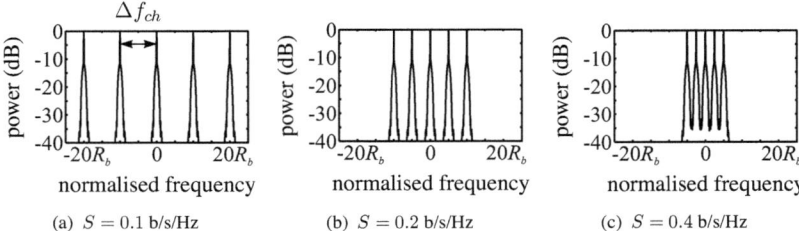

Figure 1.1: Spectrum of five WDM channels, each carrying information with bit rate R_b. Spectral efficiency is increased by reducing the channel spacing Δf_{ch}.

Therefore, a lot of research is focussed on optical amplification in other wavelength bands and on using the available bandwidth of inline optical amplifiers more efficiently, i.e. to transport more information over the same bandwidth. A measure for the information spectral density is the so called spectral efficiency

$$S = \frac{R_b}{\Delta f_{ch}}, \qquad (1.1)$$

which is the quotient of bit rate R_b per WDM channel and WDM channel spacing Δf_{ch}.

Due to cost efficiency most commercial digital fibre-optic communication systems still use the most basic form of modulating the optical carrier. The light is switched on to transmit a logical 'One' and off for a logical 'Zero'. This kind of binary modulation is referred to as on-off keying (OOK). After introduction of the erbium-doped fibre amplifier (EDFA) in the mid 1990's, which allowed broadband simultaneous amplification of many WDM channels, increased spectral efficiency could be readily achieved by reducing channel spacing and applying narrower optical bandpass filters [7]. Fig. 1.1 shows the effect of increased spectral efficiency for OOK modulated carriers in the spectral domain. For low spectral efficiency, WDM channels remain clearly separated. For increasing spectral efficiency, where channel spacing approaches the bit rate per channel, the extraction of WDM channels becomes more challenging and requires narrow-band optical filters with very steep flanks. The highest spectral efficiency reported for binary modulation is 1 b/s/Hz and was reached in 2007 [8]. Although the theoretical maximum spectral efficiency for binary modulation is 2 b/s/Hz [9], the required technological effort and investment make it infeasible to actually reach it. A further increase of spectral efficiency there-

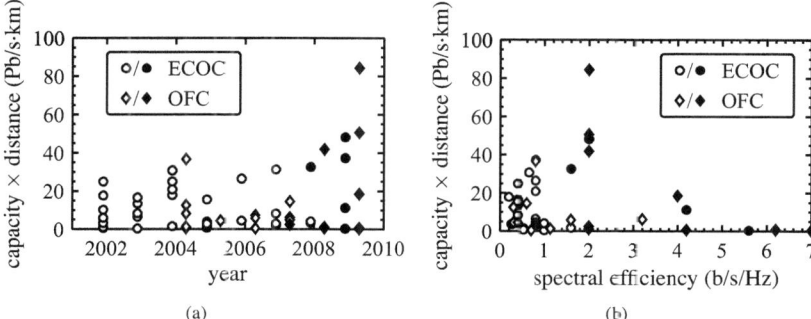

Figure 1.2: Selected experimental results reported in post-deadline sessions of recent OFCs and ECOCs. (a) Product of total capacity and distance versus year of presentation, (b) product of total capacity and distance versus achieved spectral efficiency. White symbols represent experiments using direct detection. Black symbols denote experiments with coherent detection where a digital sampling oscilloscope with subsequent off-line processing of the received data has been used.

fore necessitates the use of multilevel modulation formats, which encode more than one bit per transmitted symbol. This push to multilevel modulation and advances in electronic digital signal processing brought about a revival of coherent receivers. In recent years, phenomenal results were achieved by using multilevel modulation in conjunction with coherent detection. Fig. 1.2 summarises some selected experimental results, which were presented in the post-deadline sessions of the Optical Fiber Communication Conference (OFC) and the European Conference on Optical Communication (ECOC) over the past several years. Shown is the product of capacity times distance as a function of presentation date or spectral efficiency. White symbols denote experiments using direct detection of the incoming light, while black symbols denote coherent detection. Due to a lack of digital signal processing circuits operating at speeds of several Gb/s, all of the depicted experiments with coherent detection use a digital sampling oscilloscope and subsequent off-line processing of the data. Although first realtime transmission experiments using electronic digital signal processing have been conducted for symbol rates up to 10 GBd [10], it seems that optics has to wait for electronics to evolve and provide the required speeds. However, it becomes clear from Fig. 1.2 that multilevel modulation and coherent detection enable new record spectral efficiencies and bit rate times distance products.

Ultimately, the capacity of fibre-optic communication systems is limited by the inherent nonlinearity of silica fibre, which dictates the maximum optical launch power into the system. There has been a lot of research focussed on determination of the

3

1 Introduction

exact maximum capacity which could be transmitted over a single-mode optical fibre (e.g. [11–21]). The focus of this work is not to estimate the channel capacity but to provide a general insight into the scaling of fibre nonlinearity with important design parameters of fibre-optic communication systems. On the transmitter side, these design parameters include the choice of modulation format (including possible electronic precompensation for nonlinear effects), bit rate per WDM channel and WDM channel spacing. The nonlinear impact of the transmission line depends on the choice of a transmission fibre and dispersion compensation scheme. It is the aim of this work to develop a simplified model for fibre nonlinearity in dispersion-managed transmission systems. The main challenge here is to simplify as much as possible, while still remain applicable to many different system configurations. This is in continuation of previous work done by Hadrien Louchet [22]. The analytic framework developed by Hadrien Louchet provides a starting point for the considerations presented in the following. In particular, the idea of an equivalent single-span model is pursued. This model aims to approximate the nonlinear impact of a transmission line consisting of multiple fibre spans by an analysis of just a single fibre span, thereby significantly simplifying the optimisation process of dispersion-managed transmission systems.

Chapter 2 starts with an overview of important physical principles of signal propagation in single-mode silica fibres. It focusses on the effects of fibre loss and dispersion as well as on nonlinear interactions occurring in a single wavelength channel and nonlinear interactions between WDM channels. In chapter 3 a simplified mathematical model is derived, which approximates the nonlinear perturbation experienced by an optical signal travelling through a fibre-optic transmission line. From this simplified model, scaling laws for different nonlinear propagation effects are derived. The scaling laws are then verified by numerical simulations in chapter 4. Furthermore, the model is applied to explore the impact of multilevel modulation on the resilience of a modulation format to fibre nonlinearity. Last but not least, scaling of nonlinear effects with varying spectral efficiency is investigated.

2 Propagation Effects

This chapter briefly introduces propagation effects affecting optical signals transmitted over single-mode silica fibre. Section 2.1 deals with the linear effects of fibre loss and group-velocity dispersion (GVD) and section 2.2 with nonlinear phenomena, such as nonlinear phase modulation and mixing processes. Notation and derivations in this chapter are kept close to [23] for the most part, with exceptions being marked specifically.

The electric field of the fundamental fibre mode propagating in z-direction along a single-mode fibre can be expressed in the time domain as

$$\vec{E}(\vec{r},t) = \vec{p}F(x,y)A(z,t)\exp\left[j\left(\beta_0 z - \omega_0 t\right)\right], \tag{2.1}$$

where $\vec{r}$ is the spatial coordinate vector, $\vec{p}$ is the Jones vector describing the electric field's state of polarisation, $F(x,y)$ is the transversal field distribution of the fundamental fibre mode, $A(z,t)$ is a slowly varying (compared to the angular frequency ω_0) field envelope and β_0 is the propagation constant at ω_0. In the following, the electric field is assumed to have a constant linear state of polarisation, such that a scalar representation of the field may be used. Furthermore, the transversal field distribution is independent of propagation distance. Therefore, evolution of optical signals along the fibre can solely be described by the complex envelope $A(z,t)$.

2.1 Linear Effects

This section deals with linear propagation phenomena. When the propagated optical power is very low, a silica fibre can be approximated to be a linear transmission medium. Under the assumption of constant linear polarisation, signal distortions due to fibre birefringence (e.g. polarisation-mode dispersion) are neglected. In this case, optical signals are affected by fibre loss and GVD only, which are described in the following two sections.

2 Propagation Effects

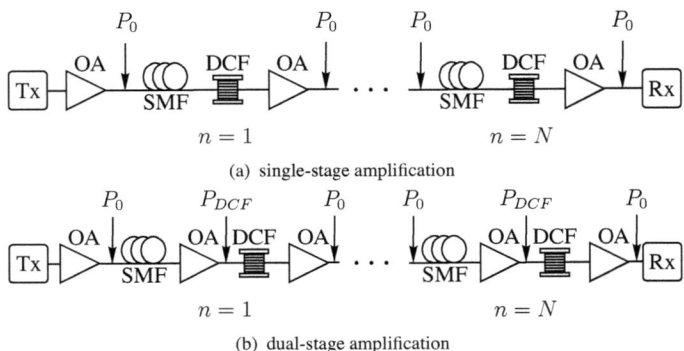

Figure 2.1: Schematic of a fibre-optic transmission system consisting of a transmitter (Tx), a receiver (Rx) and N spans. Each span comprises of a single-mode fibre (SMF), a dispersion-compensating fibre (DCF) and (a) single-stage optical amplifier (OA) or (b) dual stage OA. P_0 and P_{DCF} denote output powers of the OAs in front of SMF and DCF, respectively.

2.1.1 Fibre Loss

Light propagating along a silica fibre is attenuated through scattering and absorption processes. A more comprehensive description of the processes responsible for fibre loss can be found e.g. in [24]. The optical power P is attenuated while propagating along the fibre with

$$P(z) = P_0 \exp(-\alpha z), \qquad (2.2)$$

where P_0 is the time-averaged optical power at $z = 0$ in units W and α is the attenuation coefficient of the fibre in units Np/km. It is often more practical to express the attenuation coefficient in units dB/km. To distinguish between the two it is then referred to as fibre loss.

Typically, the fibre loss of a silica fibre is minimum at a wavelength of 1550 nm in a trade-off between Rayleigh scattering and infrared absorption. In the late 1970's, technology had evolved to fabricate low-loss fibres with measured losses of 0.2 dB/km at this wavelength [25]. Although new fibres have been developed in recent years achieving exceptionally low fibre loss (e.g. 0.1484 dB/km [26]) most field-deployed fibres still have a minimum fibre loss of about 0.2 dB/km. Therefore, unless stated otherwise, a fibre loss of $\alpha = 0.2$ dB/km is assumed throughout this thesis.

In fibre-optic transmission systems spanning distances exceeding several tenths of

2.1 Linear Effects

kilometres, the optical signal has to be amplified to compensate for fibre loss. Today there are two quite different approaches to inline amplification of WDM signals: lumped amplification using EDFAs (an extensive discussion of EDFA fundamentals can be found in [27]) and distributed amplification exploiting stimulated Raman scattering (see e.g. [28]). Both techniques are also used in combination. Although some concepts presented in this thesis also apply to Raman amplified systems, the main focus is on systems with EDFA-only amplification. A schematic of such a system is sketched in Fig. 2.1. Each EDFA compensates loss and adds amplified spontaneous emission (ASE)-noise to the signal, thereby reducing the optical signal-to-noise ratio (OSNR) [27]. The OSNR at the receiver of a transmission link with single-stage amplification according to Fig. 2.1(a) can be easily calculated (in dB) as [29]

$$\text{OSNR} = 58 + P_{in} - F_{OA} - \alpha_{SMF} L_{SMF} - \alpha_{DCF} L_{DCF} - 10 \log(N), \quad (2.3)$$

where P_{in} is the average launch power in dBm, F_{OA} the noise figure of a single optical amplifier, N the number of spans and α_{SMF}, L_{SMF}, α_{DCF}, L_{DCF} are fibre loss and length of SMF and DCF, respectively. The quantum noise limited OSNR referred to a noise bandwidth of $\Delta\lambda = 0.1$ nm is 58 dB at $P_{in} = 0$ dBm. A major drawback of this amplification scheme is that the input power into the DCF can not be adjusted individually. Therefore, a dual-stage amplification scheme as shown in Fig. 2.1(b) is usually used in practical systems. In this case, dual-stage means that there is an additional EDFA between SMF and DCF, which allows adjustment of the input power into the DCF. Furthermore, this scheme allows for a larger OSNR at the receiver due to a reduced overall noise figure of the system, when compared to systems with single-stage amplification at the same launch power. The OSNR for a system with dual-stage amplification is

$$\text{OSNR} = 58 + P_{in} - F_{DC} - \alpha_{SMF} L_{SMF} - 10 \log(N), \quad (2.4)$$

where F_{DC} is the noise figure of a dual-stage amplifier with a DCF in-between the two stages in dB. The noise figure depends on the losses of SMF and DCF as well as on their respective input powers and can be approximately calculated with [29]

$$F_{DC} = F_{OA} \left(1 + \frac{P_0 \exp(-\alpha_{SMF} L_{SMF})}{P_{DCF} \exp(-\alpha_{LCF} L_{DCF})}\right), \quad (2.5)$$

where α_{SMF} is the attenuation coefficient of the SMF, P_{DCF} is the input power into the DCF (in units Watt) and α_{DCF} and L_{DCF} are attenuation coefficient and

2 Propagation Effects

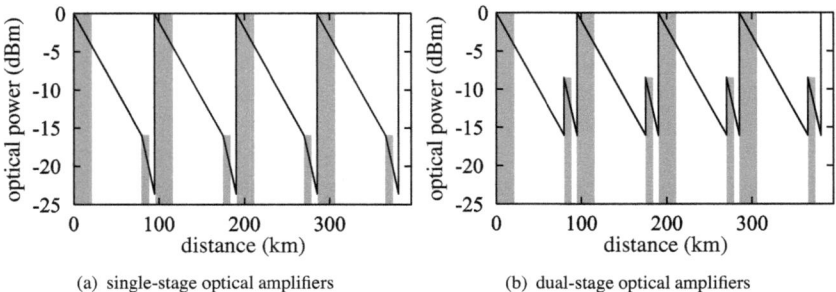

Figure 2.2: Optical power as a function of distance for transmission lines with (a) single-stage and (b) dual-stage optical amplifiers. Grey areas indicate effective lengths of SMF and DCF.

length of the DCF, respectively [1]. For example, assuming spans consisting of 100 km SMF with fibre loss $\alpha_{SMF} = 0.2$ dB/km followed by 16 km DCF with fibre loss $\alpha_{SMF} = 0.5$ dB/km and input powers $P_0 = 1$ mW and $P_{DCF} = 0.1$ mW leads with Eq. (2.5) to a noise figure of $F_{DC} = 2.1$ dB.

A useful parameter to describe the influence of fibre loss on nonlinear signal propagation is the effective length L_{eff}. It is defined as the length of a fictitious lossless fibre which has the same integrated power as the real fibre with length L and attenuation coefficient α and can be written as

$$L_{eff} = \int_0^L \frac{P(z)}{P_0} \, dz = \frac{1 - \exp(-\alpha L)}{\alpha}, \qquad (2.6)$$

with $P(z)$ according to Eq. (2.2). Fig. 2.2 shows the progression of the optical power in an exemplary transmission link consisting of four spans. While Fig. 2.2(a) plots the power in a link with single-stage EDFAs according to Fig. 2.1, Fig. 2.2(b) does the same for dual-stage EDFAs. The grey rectangles indicate the product $P_{in}L_{eff}$ for SMF and DCF ($\alpha_{DCF} = 0.5$ dB/km), respectively. As will be seen in section 2.2, the product $P_0 L_{eff}$ (with optical power P_0 in units W) is proportional to the nonlinear phase shift induced by the propagating signal. It is therefore an indicator of the strength of fibre nonlinearity in a given system configuration.

---

[1] Please note that Eq. (2.5) describes the linear noise figure, which needs to be converted to units dB for use in Eq. (2.4).

2.1 Linear Effects

2.1.2 Group-Velocity Dispersion

The refractive index of a fibre and consequently the propagation constant are a function of wavelength. For this reason, spectral components of a signal travel at different group velocities. This phenomenon is commonly referred to as group-velocity dispersion (GVD). The propagation constant can be expanded into a Taylor series around a reference angular frequency ω_0 as

$$\beta(\omega) = \beta_0 + \beta_1 (\omega - \omega_0) + \frac{1}{2}\beta_2 (\omega - \omega_0)^2 + \frac{1}{6}\beta_3 (\omega - \omega_0)^3 + \cdots, \quad (2.7)$$

where

$$\beta_m = \left.\frac{d^m \beta}{d\omega^m}\right|_{\omega=\omega_0} \quad \text{for} \quad m = 0, 1, 2, \ldots \quad (2.8)$$

The parameter $\beta_1 = 1/v_g$ describes the group-velocity v_g, $\beta_2 = d\beta_1/d\omega$ the GVD and $\beta_3 = d\beta_2/d\omega$ the second-order GVD. In this context, it should be noted that the GVD is commonly described with respect to wavelength λ rather than to angular frequency. Fibres are characterised by their dispersion parameter D in units ps/(nm·km) which relates to β_2 as

$$D = -\frac{2\pi c}{\lambda^2}\beta_2, \quad (2.9)$$

where c is the speed of light in vacuum. In the following, the dispersion parameter D is used to specify GVD values of fibres.

For linear transmission, the field envelope is described by the differential equation [23]

$$\frac{\partial A(z,T)}{\partial z} = -\frac{\alpha}{2} A(z,T) - \frac{j}{2}\beta_2 \frac{\partial^2 A(z,T)}{\partial T^2} + \frac{1}{6}\beta_3 \frac{\partial^3 A(z,T)}{\partial T^3}, \quad (2.10)$$

where $T = t - \beta_1 z$ is a retarded time frame moving with the group velocity $v_g = 1/\beta_1$. The Fourier transform of Eq. (2.10) is

$$\frac{\partial \tilde{A}(z,\omega)}{\partial z} = -\frac{1}{2}\left(\alpha - j\omega^2 \beta_2\right) \tilde{A}(z,\omega), \quad (2.11)$$

where $\tilde{A}(z,\omega)$ is the Fourier transform of $A(z,T)$ and second order GVD has been neglected. The solution of Eq. (2.11) is of the form

$$\tilde{A}(z,\omega) = \tilde{A}(0,\omega) \exp\left[-\frac{1}{2}\left(\alpha - j\omega^2 \beta_2\right) z\right]. \quad (2.12)$$

2 Propagation Effects

This means that the frequency dependence of the refractive index results in a frequency dependent phase shift of the spectral components of a signal. Although the power spectrum remains unchanged, the time-domain pulse shape becomes distorted.

Similar to the effective length, there is a length scale associated with GVD. The dispersion length is often defined as $L_D = T_0^2/|\beta_2|$, where T_0 is the initial pulse width [23]. In the context of transmission of Gaussian pulses, T_0 designates the half width at $1/e$ of peak intensity of the Gaussian pulse. In the special case of unchirped Gaussian pulses, L_D defines the distance where the pulse is broadened by a factor of $\sqrt{2}$. It is thus a simple measure for the influence of GVD on the pulse width. However, pulse shapes in practical transmission systems are not Gaussian. Examples of non-Gaussian pulse shape are return-to-zero (RZ) pulses shaped by a pulse carver based on a Mach-Zehnder interferometer [30]. In other cases, it is generally difficult to define a meaningful pulsewidth, e.g. for non-return-to-zero (NRZ) modulated signals or phase modulated signals with constant intensity. For this reason it makes sense to define the dispersion length in a more general way as

$$L_D = \frac{1}{R_s^2 |\beta_2|}, \qquad (2.13)$$

with symbol rate R_s. In this form it designates the position where two spectral components spaced $\Delta\omega = R_s$ apart, experience a group delay difference of $\Delta\tau = R_s^{-1}$ (neglecting second order GVD).

For transmission over lengths $L \gg L_D$ the cumulated dispersion has to be compensated to mitigate signal distortions induced by GVD. This task is typically performed by DCFs, which are fibres having a negative dispersion coefficient.

2.2 Nonlinear Effects

Light propagating through a silica fibre changes the refractive index of the material. This phenomenon is called optical Kerr effect and was experimentally demonstrated as early as 1973 [31].

2.2 Nonlinear Effects

Mathematically, the Kerr effect can be treated as a small perturbation to the refractive index [23]

$$n = n_l + n_{nl} = n_l + \frac{n_2}{A_{eff}} P, \qquad (2.14)$$

where n_l is the unperturbed, frequency-dependent refractive index, n_{nl} is the nonlinear perturbation of the refractive index, n_2 the nonlinear index coefficient, A_{eff} the effective core area of the fibre and P the instantaneous optical power. In measurements, the nonlinear index coefficient was found to be $n_2 = 2.6 \cdot 10^{-20}$ m²/W for standard single-mode fibre (SSMF) and $n_2 = 2.7 \cdot 10^{-20}$ m²/W for non-zero dispersion-shifted fibre (NZDSF) at a wavelength of $\lambda = 1550$ nm [32, 33]. The effective core area ranges from about 25 µm² found in DCFs to 120 µm² in specially designed large-effective area fibres [34], which are attractive due to their low nonlinearity, e. g. for transmission over transoceanic distances [35].

Analogous to the refractive index, the propagation constant becomes perturbed as

$$\beta' = \beta(\omega) + \gamma P, \qquad (2.15)$$

where $\beta(\omega)$ is the unperturbed propagation constant according to Eq. (2.7) and γ is the nonlinear coefficient of the fibre with

$$\gamma = \frac{n_2 \omega_0}{c A_{eff}}. \qquad (2.16)$$

The nonlinear coefficient is defined at the reference angular frequency $\omega_0 = 2\pi f_0$. Throughout this thesis, the reference frequency is chosen such that it is the centre frequency of the simulated frequency band at $f_0 = 193.1$ THz, which coincides with the wavelength of minimum fibre loss.

For nonlinear pulse propagation Eq. (2.10) has to be extended to include the optical Kerr effect. Neglecting GVD of second and higher orders as well as stimulated scattering processes such as stimulated Raman scattering and stimulated Brillouin scattering (a discussion of which and can be found in [23]), pulse propagation in a single-mode fibre is described by the nonlinear Schrödinger equation (NLSE) [23]

$$\frac{\partial A(z,T)}{\partial z} = -\frac{\alpha}{2} A(z,T) - j \frac{\beta_2}{2} \frac{\partial^2 A(z,T)}{\partial T^2} + j\gamma |A(z,T)|^2 A(z,T). \qquad (2.17)$$

where γ the nonlinear coefficient of the fibre.

2 Propagation Effects

The following two sections deal with nonlinear propagation effects originating in and affecting a single wavelength channel (section 2.2.1) or multiple wavelength channels (section 2.2.2).

2.2.1 Intrachannel Nonlinearities

In the late 1990's research was on the way to upgrade existing 10 Gb/s infrastructure to a bit rate of 40 Gb/s per wavelength channel (see e.g. [36–45] and references therein). In this context it became clear that new nonlinear phenomena arise, when pulses broadened by dispersion overlap with neighbouring pulses. In these high-speed systems, the dispersion length can become much smaller than the effective length of a fibre. When this is the case, many pulses overlap and interact nonlinearly with each other within the effective length. This nonlinear interaction between pulses of the same wavelength channel gives rise to the nonlinear effects intrachannel cross-phase modulation (IXPM) and intrachannel four-wave mixing (IFWM) [29,41].

A simple equation for the number of overlapping pulses has been derived in [46]. By writing this equation with the symbol rate instead of the bit rate, the number of overlapping symbols can be calculated as

$$m = C_{max} R_s \Delta\omega_s + 1, \tag{2.18}$$

where $\Delta\omega_s$ is the spectral width of the signal and C_{max} designates the maximum absolute cumulated dispersion (in units ps^2) at locations in the system with enough signal power to generate a significant nonlinear perturbation of the propagation constant [46]. For transmission over a single fibre, C_{max} corresponds to the cumulated dispersion over the effective length. It follows with Eq. (2.6) and (2.13) that the number of overlapping symbols is directly proportional to the quotient of effective length and dispersion length

$$m = \frac{L_{eff}}{L_D} \frac{\Delta\omega_s}{R_s} + 1, \tag{2.19}$$

where $\Delta\omega_s/R_s$ is a constant depending on the modulation format. It relates the spectral width $\Delta\omega_s$ of the signal to the symbol rate. For narrow-band optical filtering at the transmitter, the spectral width can approximately be taken as the bandwidth B_{MUX} of the optical multiplexer filters, i.e. $\Delta\omega_s \approx 2\pi B_{\text{MUX}}$.

2.2 Nonlinear Effects

For a small symbol rate and GVD, i.e. for $L_D \gg L_{eff}$, it can be assumed that the pulse shape does not change significantly within the effective length of a fibre. In this case, the only intrachannel nonlinearity is classical self-phase modulation (SPM) or more precisely intrapulse SPM, i.e. the nonlinear change of the refractive index depends only on the waveform of a single pulse itself. This kind of transmission regime is commonly referred to as solitonic regime as opposed to the pseudo-linear regime, where many pulses overlap during transmission [29]. The transition between these two regimes, depending on bit rate, GVD, fibre loss and modulation format has been investigated in [47]. The transition bit rate is there defined with respect to optimum net residual dispersion at the receiver. In pseudo-linear transmission, the optimum net residual dispersion is very small for NRZ-OOK (some tenths of ps/nm) and approaches zero for RZ-OOK with small duty cycle [48]. Contrary to that, the optimum net residual dispersion amounts to several hundred ps/nm in the solitonic transmission regime, where the nonlinear phase shift due to SPM can be partly compensated by residual dispersion [49]. Somewhat arbitrarily, the bit rate where the optimum net residual dispersion falls below 10% of the cumulated dispersion of a SSMF (with dispersion parameter $D = 17$ ps/(nm·km) and fibre loss $\alpha = 0.2$ dB/km) is defined as the transition bit rate in [47]. It was found that for SSMF the transition bit rates are in the ranges 20 to 22.5 Gb/s for NRZ-OOK and 17.5 to 20 Gb/s for RZ-OOK with 33% duty cycle. This corresponds to $2 < m < 3$.

Assuming three consecutive pulses of a single wavelength channel as input signal, the envelope of the electric field at the input of a fibre can be written as

$$A(0,T) = A_1(0,T) + A_2(0,T) + A_3(0,T), \tag{2.20}$$

with $A_2(0,T) = A_1(0, T - T_s)$ and $A_3(0,T) = A_1(0, T - 2T_s)$, where T_s is the symbol duration. Substituting (2.20) into the NLSE (2.17) and evaluating evolution of $A_1(z,T)$ yields (with $A_i(z,T) = A_i$ for brevity)

$$\frac{\partial A_1}{\partial z} = -\frac{\alpha}{2} A_1 - j\frac{\beta_2}{2} \frac{\partial^2 A_1}{\partial T^2} \\ + j\gamma \left(\underbrace{|A_1|^2 A_1}_{\text{(I)SPM}} + \underbrace{2|A_2|^2 A_1 + 2|A_3|^2 A_1}_{\text{IXPM}} + \underbrace{A_2^2 A_3^*}_{\text{IFWM}} \right). \tag{2.21}$$

It is obvious that as long as there is no significant pulse overlap, the contribution of IXPM and IFWM terms to this differential equation is zero (since $|A_2|^2 A_1 \approx$

13

2 Propagation Effects

$|A_3|^2 A_1 \approx A_2^2 A_3^* \approx 0$) and only intrapulse SPM remains. For pulse-overlapped transmission, all three nonlinear terms have to be accounted for.

Intrapulse Self-Phase Modulation

The phenomenon of self-phase modulation (SPM) was first measured in a silica fibre in 1978 [50]. Variation of the intensity of the electric field leads to a variation of refractive index and propagation constant, i.e. the phase of the electric field. This can be seen by solving the NLSE in the absence of dispersion. With $\beta_2 = 0$ and $\beta_3 = 0$, Eq. (2.17) reduces to

$$\frac{\partial A(z,T)}{\partial z} = -\frac{\alpha}{2} A(z,T) + j\gamma |A(z,T)|^2 A(z,T). \tag{2.22}$$

This differential equation is solved by

$$A(z,T) = A(0,T) \exp\left(-\frac{\alpha}{2} z + j\Phi_{NL}(z,T)\right), \tag{2.23}$$

with the nonlinear phase shift

$$\Phi_{NL}(z,T) = \gamma |A(0,T)|^2 \frac{1 - \exp(-\alpha z)}{\alpha}. \tag{2.24}$$

Such a time-dependent phase shift is equivalent to a frequency chirp. Leading pulse edges cause a red shift, i.e. negative chirp, while trailing pulse edges cause a blue shift, i.e. positive chirp. This leads to spectral broadening, while the pulse shape is preserved in the absence of dispersion. When the signal is transmitted at a wavelength not coinciding with the zero-dispersion wavelength, the frequency chirp is translated to amplitude distortion by phase modulation (PM)-to-amplitude modulation (AM) conversion induced by GVD [51,52]. There are special cases where this interplay between SPM and GVD can be beneficial, e.g. transmission of fundamental solitons [23]. However, WDM transmission of fundamental solitons is heavily impaired by interchannel nonlinearities and is therefore not considered to be a feasible solution for high-capacity commercial transmission systems [30].

A characteristic length scale for the SPM-induced nonlinear phase shift is the nonlinear length

$$L_{NL} = \frac{1}{\gamma P_0}. \tag{2.25}$$

2.2 Nonlinear Effects

It is here defined with respect to the time-averaged power rather than to the peak power as in [23]. With Eq. (2.6), (2.24) and (2.25), the time-averaged nonlinear phase shift at the end of a fibre of length L can be written as

$$\Phi_{NL} = \gamma P_0 \frac{1 - \exp(-\alpha L)}{\alpha} = \frac{L_{eff}}{L_{NL}}. \tag{2.26}$$

For an average nonlinear phase shift of 1 rad the nonlinear length thus equals the effective length of a fibre. Furthermore, Eq. (2.26) shows that the product $P_0 L_{eff}$, implicitly shown in Fig. 2.2 is indeed proportional to the generated nonlinear phase shift.

One implication of Eq. (2.26) is that the total SPM-induced nonlinear phase shift is larger in systems employing dual-stage optical amplifiers [Fig. 2.2(b)] compared to single-stage amplification [Fig. 2.2(a)], since the nonlinear length of the DCFs is usually smaller for dual-stage amplification. This was verified by numerical simulations in [49]. Dual-stage amplifiers allow adjustment of the input power into the DCFs. The noise figure of the transmission line can be minimised by choosing equal input powers into transmission fibre and DCF. However, due to the small effective core area of DCFs and thus large nonlinear coefficient, this would create unacceptable nonlinear distortion of the signal. Therefore, the input power into the DCF has to be optimised with respect to both noise figure and nonlinear phase shift in such systems.

Intrachannel Cross-Phase Modulation

The term intrachannel cross-phase modulation (IXPM) was first established by Essiambre in 1999 [41]. In the following years, extensive research was performed to understand the basic principles and system implications of IXPM. First theoretical studies analysed frequency and timing shift induced by IXPM for pulse-overlapped transmission of two Gaussian pulses for single-fibre [53,54] and dispersion-managed transmission [55]. These results were soon generalised to transmission of random sequences of Gaussian pulses and the generation of timing jitter observed in experiments was theoretically described for systems with dispersion compensation at the receiver [56] and periodic dispersion-management [57,58].

2 Propagation Effects

The principle of IXPM can be demonstrated with two Gaussian pulses associated with the same wavelength channel. While the pulses remain separated in the time domain there is no nonlinear interaction between them. However, when the pulses acquire dispersion they are broadened and overlap eventually [see inset of Fig. 2.3(b)]. In this case, IXPM terms in (2.21) contribute to the total nonlinear phase shift experienced by pulse A_1. An analytic expression for the IXPM induced chirp depending on the amount of pulse broadening of two Gaussian pulses was derived in [53]. It was found that the chirp initially increases and is maximum when the full width at half maximum of the pulses approaches the initial pulse separation. When the pulses are broadened further, the chirp is reduced, since it is proportional to the slope of the pulse edge, which decreases with pulse broadening.

Transmission of random sequences of highly dispersed pulses leads to generation of timing jitter. Each pulse is chirped by a random number of overlapping pulses. Due to GVD this random chirp leads to statistic variations of the group delay. An analytical expression for the standard deviation of the pulse arrival time at the receiver and for the optimum amount of dispersion precompensation to minimise IXPM induced timing jitter for the case of Gaussian pulses can be found in [56]. A numerical study of the reduction of IXPM by optimisation of the dispersion precompensation was performed in [59]. Apart from dispersion-management, IXPM can also be reduced by alternating the state of polarisation between two orthogonal states from bit to bit, since nonlinear interaction between orthogonally polarised signal components is reduced by 2/3 [60]. Optical phase conjugation, originally proposed to compensate for SPM [61] in the solitonic regime, is also an effective means to combat IXPM in pseudo-linear transmission as was shown in an experiment for differential phase-shift keying (DPSK) in [62].

Intrachannel Four-Wave Mixing

The phenomenon of intrachannel four-wave mixing (IFWM) was first measured in a 100 Gb/s optical time-division multiplexing (OTDM) transmission experiment in 1998 [63]. In OOK transmission, IFWM leads to the generation of so called ghost pulses at the temporal positions where logical zeroes are transmitted. Later, IFWM was also experimentally confirmed for transmission of 40 Gb/s electrical time-division multiplexing (ETDM) signals with RZ pulse shaping over SSMF [41]. In the following years, the generation of ghost pulses was analytically described

2.2 Nonlinear Effects

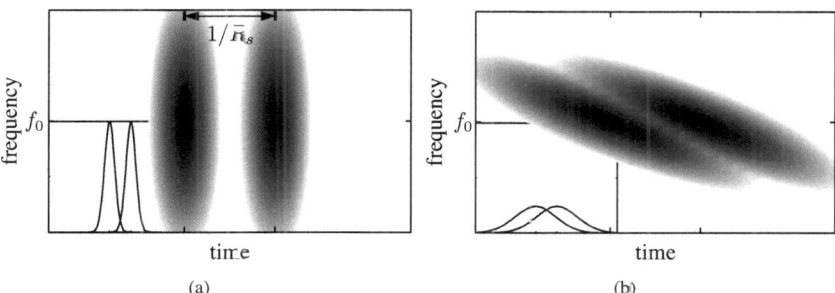

Figure 2.3: Spectrogram (in logarithmic scale) of two Gaussian pulses (a) at the input of the fibre and (b) broadened by GVD at $z = 4L_D$. Insets show pulse shapes in the time domain.

using perturbation theory for the exemplary case of Gaussian pulse shape and constant dispersion [54] as well as periodic dispersion [64]. In later studies, it was found that in systems with small average dispersion, IFWM induced energy transfer between pulses and therewith amplitude jitter and amplitude of ghost pulses grow linearly with transmission distance [58, 65–67].

In order to illustrate the origin of IFWM, it is convenient to use a spectrogram representation of the optical signal. A spectrogram of two Gaussian pulses computed with short-time Fourier transform [68] using a sliding Gaussian window of width $0.7T_0$ is depicted in Fig. 2.3(a). When dispersion accumulates, different group-velocities of red and blue spectral components lead to walk-off between them, which can be seen as a tilt of the spectrogram [Fig. 2.3(b)], and the pulses eventually overlap. Just as with light from different wavelength channels, nonlinear mixing occurs between the red spectral components of the leading pulse and the blue spectral components of the trailing pulse ($\beta_2 < 0$ assumed). The frequencies of the mixing products and the efficiency of this four-wave mixing (FWM) process are covered in detail in section 2.2.2. After dispersion compensation, the generated mixing products emerge as ghost pulses in the time domain. For the example of three consecutive pulses in Eq. (2.20), the mixing products are reformed around time instants [64]

$$T_{ijk} = T_i + T_j - T_k \quad \text{with} \quad i,j \neq k, \tag{2.27}$$

where T_i, T_j and T_k are the centres of the respective symbol time slots. An example for two Gaussian pulses can be found in [69].

17

For high-speed transmission systems, the generation of ghost pulses by IFWM is one of the most detrimental effects [29]. A lot of research has been focussed on how to reduce the impact of IFWM on transmission performance. In the same way as IXPM, IFWM can be reduced by proper dispersion-management [59] and by employing bit-to-bit alternate polarisation with OOK [60, 70–72] and DPSK formats [73–78]. Unlike the incoherent process IXPM, IFWM is susceptible to additional phase coding of the signal pulses. It has been shown that proper phase coding can tremendously reduce the build-up of ghost pulses in OOK transmission [79–85]. A further reduction of IFWM can be achieved with a combination of both polarisation and phase coding techniques by introducing additional phase coding in each of the two orthogonal states of polarisation [86, 87].

2.2.2 Interchannel Nonlinearities

The term interchannel nonlinearities comprises all nonlinear interactions between different copropagating wavelength channels of a WDM signal. There are two distinguished scalar interchannel nonlinear effects: cross-phase modulation (XPM) and four-wave mixing (FWM). Very detailed analyses of these effects can be found e.g. in [23, 88, 89]. In the following two sections, a brief description of the main features and system impact of XPM and FWM is given.

For the discussion of interchannel nonlinear effects, it is useful to rewrite the NLSE for a WDM-signal in the limit of vanishing dispersion. Transmission of three wavelength channels with field envelopes $A_1(z,T)$, $A_2(z,T)$ and $A_3(z,T)$ is considered. The carrier frequencies of the WDM channels are located at angular frequencies ω_1, ω_2 and ω_3 and are spaced $\Delta\omega_{ch}$ apart. In this case, the field envelope of the propagating electric field can be written as a superposition

$$A(z,T) = A_1(z,T) + A_2(z,T) + A_3(z,T). \tag{2.28}$$

Substituting Eq. (2.28) into Eq. (2.22) and evaluating all terms forming at ω_1 yields (with $A_i = A_i(z,T)$ for brevity)

$$\frac{\partial A_1}{\partial z} = -\frac{\alpha}{2}A_1 + j\gamma \left(\underbrace{|A_1|^2 A_1}_{\text{SPM}} + \underbrace{2|A_2|^2 A_1 + 2|A_3|^2 A_1}_{\text{XPM}} + \underbrace{A_2^2 A_3^*}_{\text{FWM}} \right). \tag{2.29}$$

2.2 Nonlinear Effects

This equation describes the evolution of the wavelength channel centred at angular frequency ω_1. Similar equations result for other wavelength channels. The SPM-term accounts for all intrachannel nonlinearities, which were described in the previous section. The influence of the other nonlinear terms responsible for XPM and FWM is discussed in the next two sections.

Cross-Phase Modulation

As the name suggests, XPM leads to a nonlinear phase shift similar to SPM. In this case, however, the nonlinear phase shift originates in the power propagated in other wavelength channels as can be seen in Eq. (2.29). By assuming negligible GVD and applying the same methodology as used in section 2.2.1 for SPM, it can be shown that the time-averaged nonlinear phase shift experienced by a single wavelength channel at the end of a fibre of length L is

$$\Phi_{NL} = \frac{L_{eff}}{L_{NL}} \left(\underbrace{1}_{\text{SPM}} + \underbrace{2N_{ch}}_{\text{XPM}} \right), \tag{2.30}$$

where N_{ch} is the number of WDM channels and equal power per channel has been assumed. Similar to SPM, XPM leads to spectral broadening. Depending on the relative timing of the interfering pulses, this broadening can be asymmetric [23,90]. Only in conjunction with nonnegligible dispersion the XPM-induced frequency chirp leads to signal distortions in the time domain.

The effect of XPM was first studied in the context of phase modulated signals, where XPM-induced phase noise degrades the signal [91,92]. Later work focussed on intensity modulated signals, particularly Gaussian pulses [90]. The nonlinear phase shift itself does not depend on spacing between WDM channels. However, two pulses belonging to different wavelength channels and being aligned with each other at the input of the fiber will walk-off due to GVD. The length over which this walk-off occurs depends on channel spacing and GVD. This characteristic length is called walk-off length and was defined in [90] as $L_w = T_0 / \left| v_{g1}^{-1} - v_{g2}^{-1} \right|$, where T_0 is the pulse width and v_{g1} and v_{g2} are the group-velocities of the two interacting pulses. The walk-off length can be approximated with good accuracy to be $L_w \approx T_0/(|\beta_2| \Delta\omega_{ch})$, i.e. it is inversely proportional to channel spacing. In some cases the walk-off length is not defined by the pulse width but rather the transition time from zero intensity

2 Propagation Effects

to peak intensity, e.g. in [93,94]. This takes into account that frequency chirp is only generated at the pulse edges. In order to be consistent with the definition of the dispersion length in Eq. (2.13) the walk-off length is here defined with respect to symbol rate

$$L_w = \frac{1}{|\beta_2| R_s \Delta\omega_{ch}}, \tag{2.31}$$

such that it designates the length after which two wavelength channels, spaced $\Delta\omega_{ch}$ apart, experience a walk-off of one symbol period. This definition is more general as it is not only applicable for pulsed signals but also for NRZ-type and constant-intensity signals. The dependence of XPM on the number of copropagating WDM channels and channel spacing, i.e. the effect of walk-off between wavelength channels, is covered in detail in [95]. It was shown that for $L_w \ll L_{eff}$ the signal distortion caused by XPM is reduced and performance of the WDM system approaches that of a single-channel system [93,95]. This can also be understood theoretically by looking at the efficiency of XPM derived in [96], which becomes proportional to L_w^2 for $L_w \ll L_{eff}$.

It is intuitively obvious that there exists a trade-off between PM-to-AM conversion by GVD and the benefit of having a smaller walk-off length. Indeed, it was found that the XPM-induced penalty for WDM transmission of 10 Gb/s NRZ-OOK signals is smaller when transmitting at the zero-dispersion wavelength, e.g. over dispersion-shifted fibre (DSF), compared to transmission over fibres with a small amount of GVD. A further increase of GVD significantly reduces the observed penalty as the benefits of walk-off outweigh a larger efficiency of PM-to-AM conversion [97]. In another study, it was shown that the effect of XPM can be described by two parameters, namely the XPM-induced phase shift and normalised fibre dispersion DLR_s^2 [98]. For constant channel spacing, the normalised dispersion characterises the walk-off between the WDM channels and the amount of PM-to-AM conversion.

Four-Wave Mixing

First measurements of four-wave mixing (FWM) in silica fibres were performed with multi-mode fibres in 1974 [99] and with single-mode fibres four years later [100]. It was found that the optical Kerr effect leads to frequency mixing between copropagating waves and consequently to the generation of mixing products in the frequency domain. Due to the optical Kerr effect originating in the third-order susceptibility of silica, three waves mix with each other (therefore FWM is often also referred to as three wave mixing) to generate a new fourth wave. Assuming an electric field consisting of three wavelength channels as in Eq. (2.28), substituting into Eq. (2.22) and evaluating the term describing the optical Kerr effect results in the following equation for the angular frequency of the generated mixing products

$$\omega_{ijk} = \omega_i + \omega_j - \omega_k \quad \text{with} \quad i, j \neq k. \tag{2.32}$$

When employing equally spaced wavelength channels (as it is recommended e. g. by the standardisation group of the International Telecommunication Union (ITU), [101, 102]), some of these mixing products fall into the signal bandwidth of other WDM channels resulting in nonlinear crosstalk. This crosstalk as well as FWM-induced excess loss of signal power is a major source of degradation in WDM transmission systems and limits the possible input power per channel [103].

In order to occur with sufficient efficiency, the process of FWM depends on phase-matching of the interacting waves [104]. The power of the mixing products at the end of a fibre with length L is approximately determined by [88]

$$P_{ijk} = \left(\frac{d_{ijk}}{3}\gamma L_{eff}\right)^2 P_i P_j P_k \exp(-\alpha L) |\eta|^2, \tag{2.33}$$

where P_i, P_j and P_k are the input powers of the interacting wavelength channels, $|\eta|^2$ is the FWM efficiency describing phase matching and d_{ijk} is the degeneracy factor with $d_{ijk} = 3$ for $i = j$ (degenerate FWM) and $d_{ijk} = 6$ for $i \neq j$. Neglecting the influence of GVD-slope (which is legitimate when operating away from the zero-dispersion wavelength) and assuming a fibre with length $L \gg L_{eff}$, the FWM efficiency is [104]

$$|\eta|^2 = \frac{1}{1 + \left(\frac{\beta_2 \Delta\Omega}{\alpha}\right)^2}, \tag{2.34}$$

where $\Delta\Omega = (\omega_i - \omega_k)(\omega_j - \omega_k)$ describes the frequency mismatch of the interacting waves. Since $\Delta\Omega$ is proportional to the square of channel spacing and FWM efficiency is inversely proportional to the square of $\Delta\Omega$, FWM-induced distortions are reduced the farther away interacting WDM channels are located. Therefore, the input power limitation, imposed on a system by FWM, saturates for a certain number of WDM channels. This number is determined by GVD, fibre loss and channel spacing [105].

2 Propagation Effects

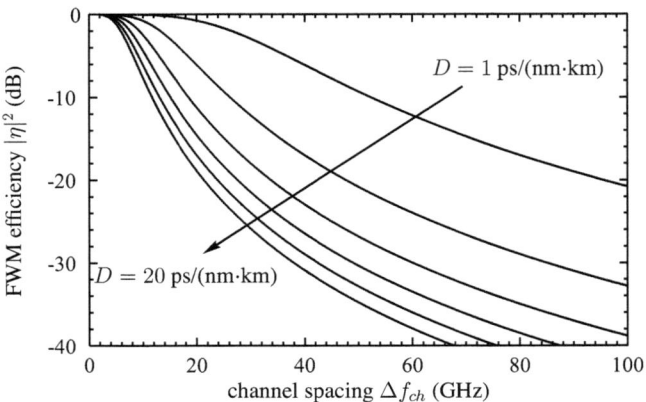

Figure 2.4: FWM efficiency as a function of channel spacing Δf_{ch} for transmission of two wavelength channels over 80 km fibre with fibre loss $\alpha = 0.2$ dB/km and varying dispersion parameter. The curves correspond to dispersion parameters $D = 1, 4, 8, 12, 16$ and 20 ps/(nm·km) at $\lambda = 1.55$ µm.

Fig. 2.4 shows the decay of FWM efficiency with frequency separation Δf_{ch} of neighbouring WDM channels (with $\Delta \Omega = 4\pi^2 \Delta f_{ch}^2$) for transmission over 80 km fibre. The dispersion parameter of the fibre is varied from 1 ps/(nm·km) to 20 ps/(nm·km). Comparison of the curves for different dispersion parameters demonstrates the effect of GVD on phase matching. As can be deduced from Eq. (2.34), the 3-dB bandwidth of FWM efficiency is inversely proportional to β_2 and the efficiency thus decays more rapidly for larger dispersion.

Apart from employing fibres with large GVD, there were other proposals how to combat FWM. One idea was to use unequal channel spacings and arrange the wavelength channels such that the FWM products do not fall into signal bands but rather in-between the wavelength channels [106]. However, determining the correct wavelengths turned out to become a very complex problem for systems with many WDM channels and conflicts with the aim of standardised channel spacings as it is pursued e.g. by the ITU [101,102]. Another idea was to design special dispersion-distributed fibres, which can be tailored to significantly reduce phase-matching [107]. However, compared to SSMF the manufacturing process of such fibres is very complex and costly.

3 Single-Span Model

It is the aim of this work to find simple, universal scaling laws for the seemingly very different nonlinear effects discussed in the previous chapter. So far, the discussion of nonlinear effects was focussed on propagation through a single fibre. However, practical fibre-optic transmission systems consist of multiple concatenated fibres and usually incorporate optical amplifiers and means of dispersion compensation. Furthermore, there exist numerous fibre types and modulation formats, each with their very own strengths and weaknesses. Numerical optimisation of the impact of nonlinear effects in this huge parameter space can become a very cumbersome and time-consuming task. It is therefore of great interest to develop simple analytical design rules guiding system engineers to the most promising solutions.

Since the design rules ought to be as simple as possible, the main challenge will be the simplification of the complex analytical description of nonlinear propagation in optical fibres. Furthermore, the developed scaling laws should be valid for as many different system configurations as possible. To tackle this task, the idea of an equivalent single-span model – briefly introduced in the work by Hadrien Louchet [22] – is pursued. This model aims to approximate the nonlinear perturbation of the received signal after propagation along a transmission line consisting of multiple fibre spans by a nonlinear perturbation generated in a single fibre span. This approach could potentially simplify the optimisation of fibre-optic transmission systems, since a single-span optimisation would yield valuable information about optimum system parameters of multi-span systems.

For the derivation of the single-span model, a Volterra series expansion[1] is applied. The Volterra series expansion as a tool to model nonlinear systems has been applied to optical fibres by Peddanarappagari and Brandt-Pearce in 1997 [108]. In 1998 the model was extended to the more general case of transmission lies consisting of amplified fibre sections [109]. Since then, it has successfully been used to

---

[1]The Volterra series expansion is named after its creator Vito Volterra, who developed this series expansion in 1887.

3 Single-Span Model

model interchannel nonlinear effects such as XPM and FWM [110]. More recently, a time-domain counterpart of the frequency-domain Volterra transfer function – the power-weighted dispersion distribution function – was applied to analytically derive the generation of ghost pulses due to IFWM in high-speed pulse-overlapped transmission [111]. The major drawback of the Volterra series expansion is its slow convergence and insufficient accuracy for large input powers, i.e. when operating in the highly nonlinear regime [112]. However, its accuracy is comparable to other alternative methods [113].

Proper dispersion management can reduce the accumulation of nonlinear distortion in fibre-optic transmission systems. Consequently, Hadrien Louchet applied the Volterra series to systems with single-periodic dispersion maps and derived a closed form solution for the received optical signal. However, the described equivalent single-fibre model does not account for dispersion precompensation [22, 114]. Here, this approach is extended to a more general class of systems, including dispersion precompensation and randomly varying residual dispersion per span (RDPS). Furthermore, an approximation of the third-order Volterra kernel transform is proposed in order to yield an equivalent single-span description with simple parameters to characterise the nonlinear properties of a wide range of possible system configurations.

Section 3.1 presents a brief introduction to nonlinear system theory and one of its mathematical tools, the Volterra series expansion. In section 3.2, the nonlinear transfer function (NLTF), which links the input electric field to the output nonlinear perturbation, is derived for a single fibre and a concatenation multiple fibre sections. Subsequently, the NLTF is simplified for transmission links consisting of identical spans with periodic dispersion-management. Finally, a randomly varying residual dispersion as it is common in practical dispersion-managed transmission systems is also considered. Some parts of this and the following chapter are taken from [115].

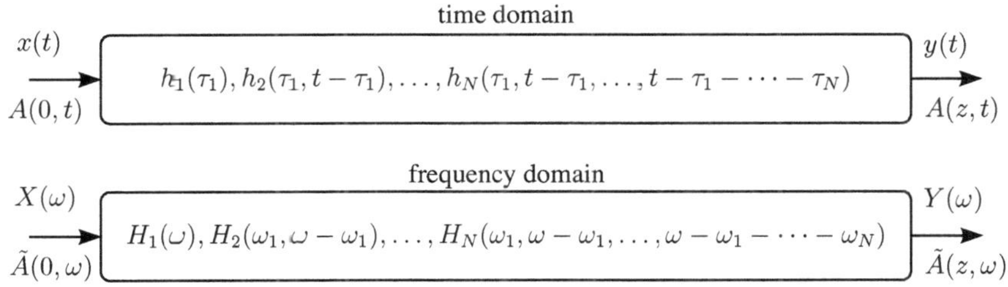

3.1 Volterra Series Expansion

time domain

$x(t)$, $A(0,t)$ → $h_1(\tau_1), h_2(\tau_1, t - \tau_1), \ldots, h_N(\tau_1, t - \tau_1, \ldots, t - \tau_1 - \cdots - \tau_N)$ → $y(t)$, $A(z,t)$

frequency domain

$X(\omega)$, $\tilde{A}(0,\omega)$ → $H_1(\omega), H_2(\omega_1, \omega - \omega_1), \ldots, H_N(\omega_1, \omega - \omega_1, \ldots, \omega - \omega_1 - \cdots - \omega_N)$ → $Y(\omega)$, $\tilde{A}(z,\omega)$

Figure 3.1: Block diagram of the optical fibre modelled as a nonlinear time-invariant system. Time-domain inputs and outputs are linked through Volterra kernels $h_1, \ldots, h_N$ and frequency-domain inputs and outputs by kernel transforms $H_1, \ldots, H_N$.

3.1 Volterra Series Expansion

In a time-invariant nonlinear system with memory, the Volterra series describes the time-domain output of the system, $y(t)$, by a sum of convolutions as [116]

$$y(t) = \int_{-\infty}^{\infty} h_1(\tau_1) x(t - \tau_1) \, d\tau_1 \\ + \iint_{-\infty}^{\infty} h_2(\tau_2) x(t - \tau_1) x(t - \tau_1 - \tau_2) \, d\tau_1 \, d\tau_2 + \ldots, \quad (3.1)$$

where $x(t)$ is the input of the system and h_n is the nth order Volterra kernel. A first-order Volterra series is easily recognised to describe a linear time-invariant (LTI) system. In this case, the first order Volterra kernel $h_1(t)$ is the impulse response of the LTI system and convolution with the input signal $x(t)$ describes the output of the system. Analogous to the relationship of time-domain impulse response and frequency-domain transfer function in LTI systems, Eq. (3.1) can be Fourier transformed to yield a relation between frequency-domain input and output as [116]

$$Y(\omega) = H_1(\omega) X(\omega) \\ + \int_{-\infty}^{\infty} H_2(\omega_1, \omega - \omega_1) X(\omega_1) X(\omega - \omega_1) \, d\omega_1 \\ + \iint_{-\infty}^{\infty} H_3(\omega_1, \omega - \omega_1, \omega - \omega_1 - \omega_2) \\ \times X(\omega_1) X(\omega - \omega_1) X(\omega - \omega_1 - \omega_2) \, d\omega_1 \, d\omega_2 + \ldots, \quad (3.2)$$

where $X(\omega)$ and $Y(\omega)$ are the Fourier transforms of $x(t)$ and $y(t)$, respectively, and H_n is the nth order kernel transform.

3.2 Nonlinear Transfer Function

Propagation of the complex envelope $A(T, z)$ of the electric field along the spatial coordinate z of a single-mode fibre is described by the NLSE. Neglecting higher-order dispersion and stimulated inelastic scattering processes, such as stimulated Raman scattering, the NLSE can be expressed in the time domain as in Eq. (2.17). In order to derive the frequency-domain Volterra series it is useful to transform Eq. (2.17) into the frequency domain. The Fourier transform of Eq. (2.17) is

$$\begin{aligned}\frac{\partial \tilde{A}(\omega, z)}{\partial z} &= -\frac{\alpha}{2}\tilde{A}(\omega, z) + j\omega^2\frac{\beta_2}{2}\tilde{A}(\omega, z) \\ &+ j\gamma \iint_{-\infty}^{\infty} \tilde{A}(\omega_1, z)\tilde{A}^*(\omega_2, z)\tilde{A}(\omega_3, z)\,\mathrm{d}\omega_1\,\mathrm{d}\omega_2,\end{aligned} \quad (3.3)$$

where $\tilde{A}(\omega, z)$ is the Fourier transform of $A(T, z)$ and $\omega_3 = \omega - \omega_1 + \omega_2$ for brevity. In a weakly nonlinear regime, the complex envelope $\tilde{A}(\omega, z)$ can be approximated with a Volterra series expansion up to third order [108]. It follows with Eq. (3.2) and the identities $\tilde{A}(\omega, z = 0) \equiv X(\omega)$ and $\tilde{A}(\omega, z) \equiv Y(\omega)$ that

$$\begin{aligned}\tilde{A}(\omega, z) &\approx H_1(\omega, z)\,\tilde{A}_0(\omega) + \iint_{-\infty}^{\infty} H_3(\omega_1, \omega_2, \omega, z) \\ &\quad \times \tilde{A}_0(\omega_1)\,\tilde{A}_0^*(\omega_2)\,\tilde{A}_0(\omega_3)\,\mathrm{d}\omega_1\,\mathrm{d}\omega_2,\end{aligned} \quad (3.4)$$

where $\tilde{A}_0(\omega)$ stands for $\tilde{A}(\omega, z = 0)$ and $H_1(\omega, z)$ and $H_3(\omega_1, \omega_2, \omega, z)$ are the first- and third-order Volterra kernel transforms, respectively. In the following

$$\tilde{S}_0 = \tilde{A}_0(\omega_1)\,\tilde{A}_0^*(\omega_2)\,\tilde{A}_0(\omega_3) \quad (3.5)$$

will be used for conciseness of notation.

3.2.1 Nonlinear Transfer Function of a Single Fibre

The Volterra kernel transforms can be determined by substituting Eq. (3.4) into Eq. (3.3). Discarding all terms of higher order than three and comparing the terms of equal order yields two differential equations for the Volterra kernel transforms $H_1(\omega, z)$ and $H_3(\omega_1, \omega_2, \omega, z)$ [108]. These differential equations are solved by

$$H_1(\omega, z) = e^{\left(-\frac{\alpha}{2} + j\omega^2 \frac{\beta_2}{2}\right)z} \qquad (3.6)$$

and

$$H_3(\omega_1, \omega_2, \omega, z) = j\gamma H_1(\omega, z) \int_0^z e^{(-\alpha + j\beta_2 \Delta\Omega)z'} \, dz', \qquad (3.7)$$

where

$$\Delta\Omega = (\omega - \omega_1)(\omega_2 - \omega_1). \qquad (3.8)$$

Substitution of Eq. (3.6) and (3.7) in (3.4) yields the field envelope after propagation

$$\tilde{A}(\omega, z) \approx H_1(\omega, z) \left(\tilde{A}_0(\omega) + \tilde{\delta}_{NL}(\omega, z) \right), \qquad (3.9)$$

where

$$\tilde{\delta}_{NL}(\omega, z) = j\gamma \iint_{-\infty}^{\infty} \int_0^z e^{(-\alpha + j\beta_2 \Delta\Omega z')} \, dz' \, \tilde{S}_0 \, d\omega_1 \, d\omega_2 \qquad (3.10)$$

is the first-order nonlinear perturbation [14, 113, 114].

Examination of Eq. (3.9) reveals that energy conservation is seriously violated when the energy contained in the nonlinear perturbation gets too large. Therefore Eq. (3.9) is valid in a weakly nonlinear regime only. This is covered in greater detail in [112]. One example given in [112] states that the pulse energies of Gaussian pulses diverge up to 10% after transmission over 100 km fibre with GVD $\beta_2 = 2$ ps²/nm, fibre loss $\alpha = 0.2$ dB/km, nonlinear coefficient $\gamma = 2$ W⁻¹km⁻¹ and input peak power of 10 mW. In the following, the nonlinear distortions are assumed small enough, such that the condition of a weakly nonlinear regime is fulfilled.

There are two distinct contributions to the nonlinear perturbation in Eq. (3.10), one stemming from the physical parameters of the fiber and the other from the input signal $\tilde{S}_0$. Thus it makes sense to describe the signal-independent part of the nonlinear perturbation with a nonlinear transfer function (NLTF). Evaluation of the inner

3 Single-Span Model

integral of Eq. (3.10) yields the NLTF as

$$\eta(\Delta\Omega, z) = \gamma \int_0^z e^{(-\alpha + j\beta_2 \Delta\Omega)z'} \, dz' \\
= \frac{\gamma}{\alpha} \frac{1 - e^{(-\alpha + j\beta_2 \Delta\Omega)z}}{1 - j\frac{\beta_2}{\alpha}\Delta\Omega}. \quad (3.11)$$

Now, the nonlinear perturbation can be concisely written as

$$\delta_{NL}(\omega, z) = j \iint_{-\infty}^{\infty} \eta(\Delta\Omega, z) \, \tilde{S}_0 \, d\omega_1 \, d\omega_2. \quad (3.12)$$

An analogue expression for the kernel in Eq. (3.11) has been derived through averaged propagation models in earlier contributions by Turitsyn and coworkers [117, 118].

For spatial coordinates $z \gg L_{eff}$, the nonlinear transfer function $\eta(\Delta\Omega, z)$ can be approximated to be independent of z. The nonlinear transfer function of such a long fibre is then obtained from Eq. (3.11) as

$$\eta_s(\Delta\Omega) \approx \frac{\gamma}{\alpha} \frac{1}{1 - j\mathrm{sgn}(\beta_2) \frac{\Delta\Omega}{\Omega_s}}, \quad (3.13)$$

where

$$\Omega_s = \alpha / |\beta_2| \quad (3.14)$$

is its 3 dB-bandwidth[2] and $\mathrm{sgn}(x)$ is the sign function defined as

$$\mathrm{sgn}(x) = \begin{cases} -1 & \text{for } x < 0 \\ 0 & \text{for } x = 0 \\ 1 & \text{for } x > 0. \end{cases} \quad (3.15)$$

Eq. (3.13) is closely related to the theory of FWM, where $\beta_2 \Delta\Omega$ describes the phase mismatch of the interacting fields and $|\eta_s|^2$ is proportional to the efficiency of the FWM process, according to Eq. (2.34). Fig. 3.2 shows the magnitude and phase of the NLTF as a function of $\Delta\Omega$ for 80 km of NZDSF with a dispersion parameter of $D = 4$ ps/(nm·km) and SSMF with $D = 16$ ps/(nm·km). For this length of fibre,

---

[2]The 3 dB-bandwidth is related through $\Omega_s = 4\pi f_d^2$ to the nonlinear diffusion bandwidth derived in [114].

3.2 Nonlinear Transfer Function

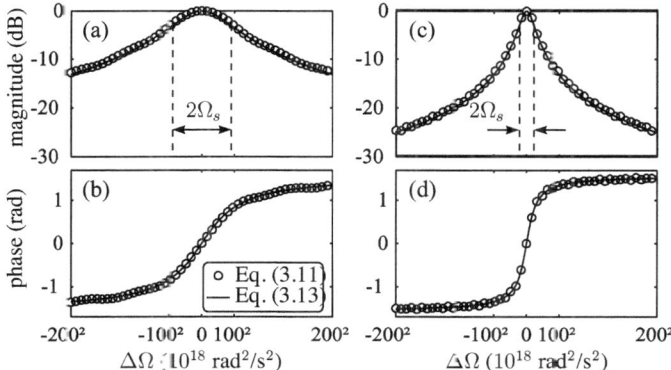

Figure 3.2: Normalised magnitude and phase of the nonlinear transfer function $\eta_s(\Delta\Omega)$ of a single fibre with $L = 80$ km, $\alpha = 0.2$ dB/km and (a,b) $D = 4$ ps/(nm·km), i.e. $\Omega_s = 9 \cdot 10^{21}$ rad^2/s^2, and (c,d) $D = 16$ ps/(nm·km), i.e. $\Omega_s = 2.25 \cdot 10^{21}$ rad^2/s^2, at a wavelength of $\lambda = 1.55$ µm. Shown are the exact function according to Eq. (3.11) (circles) and its approximation according to Eq. (3.13) (line).

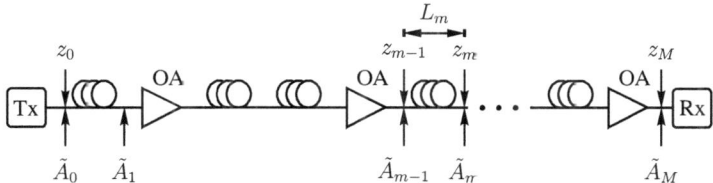

Figure 3.3: Fibre-optic transmission line consisting of M fibre sections and optical amplifiers.

the approximation of Eq. (3.11) by Eq. (3.13) is already very good. Comparing Fig. 3.2(a) and (b), it can be seen that lower local dispersion leads to a broadening of the transfer function. Consequently, more frequency components of the input signal participate significantly in the nonlinear processes affecting the signal at angular frequency ω. This reflects the well-known fact, that the impact of FWM is more severe at lower local dispersion [30, 89].

3 Single-Span Model

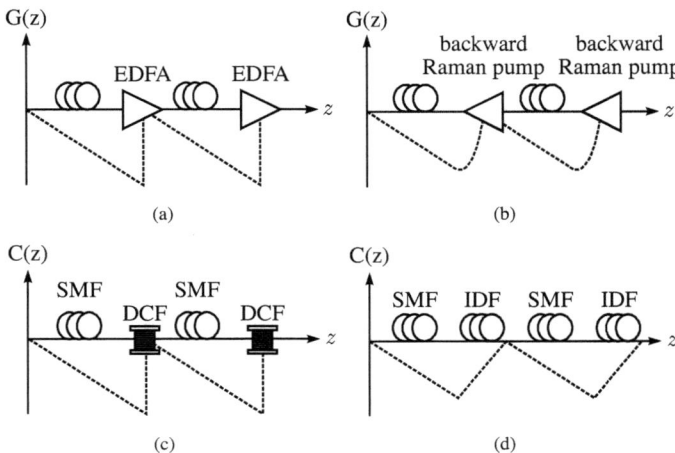

Figure 3.4: (a) Cumulated gain for lumped amplification using EDFAs at positions z_i, (b) cumulated gain for distributed Raman amplification using backward pumping, (c) cumulated dispersion for SMF spans with full inline dispersion compensation using DCFs approximated as ideal lumped dispersion-compensating devices at positions z_j and (d) cumulated dispersion for SMF spans with full inline dispersion compensation using inverse dispersion fibre (IDF).

3.2.2 Nonlinear Transfer Function of Multiple Fibre Sections

Practical fibre-optic transmission systems consist of several concatenated fibre sections. A schematic of such a system is shown in Fig. 3.3. The physical parameters of the fibres, such as length, group-velocity dispersion, attenuation and nonlinear coefficient are likely to be different in each section. Furthermore, the signal has to be amplified in certain intervals. This can either be done between fibre sections with discrete amplifiers such as e.g. EDFAs or distributed along the fibre using a Raman amplification scheme [28]. Peddanarappagari and Brandt-Pearce derived a solution for the field envelope after transmission over several fibre sections including lumped amplifiers and amplifier noise in [109]. However, due to its complexity it is quite unwieldy and does not lend itself easily to analytic examination. Approximations in the following derivation serve the aim to considerably simplify the expressions describing the field envelope after transmission over several fibre sections.

3.2 Nonlinear Transfer Function

To derive the NLTF of a system consisting of multiple fibre sections and amplifiers, including different amplification and dispersion compensation schemes, it is useful to define gain and dispersion profiles of the transmission link. They are a convenient way to describe the z-dependence of attenuation (or gain) and GVD. The gain profile can be defined as [114]

$$\frac{\mathrm{d}G(z)}{\mathrm{d}z} = -\alpha(z) + g(z) + \sum_i g_i \delta(z - z_i), \qquad (3.16)$$

where $G(z)$ is the cumulated gain[3], $g(z)$ accounts for distributed amplification, g_i is the gain of lumped amplifiers positioned at z_i and $\delta(z)$ is Dirac's delta function. Similarly, the dispersion profile can be defined as [114]

$$\frac{\mathrm{d}C(z)}{\mathrm{d}z} = \beta_2(z) + \sum_j C_j \delta(z - z_j), \qquad (3.17)$$

where $C(z)$ is the cumulated dispersion at position z and C_j is the cumulated dispersion of a lumped dispersion compensation module at position z_j. The first- and second-order Volterra kernel transforms defined in Eq. (3.6) and (3.7) depend on gain and dispersion profile, which substitute attenuation coefficient α and GVD β_2. Fig. 3.4 schematically shows cumulated gain and dispersion for some exemplary system configurations.

Fig. 3.5 shows the block diagram of a single fibre section preceded by an optical amplifier. The input field $\tilde{A}_{m-1}(\omega)$ is amplified according to the transfer function $H_a(\omega)$ of the optical amplifier, which also adds noise $\tilde{n}_m(\omega)$ to the signal. The resulting field $\tilde{A}_{m-1}(\omega)H_a(\omega) + \tilde{n}_m(\omega)$ then propagates over the fibre with associated kernel transforms $H_{1,m}(\omega)$ and $H_{3,m}(\omega)$. The output field $\tilde{A}_m(\omega)$ is determined with Eq. 3.4, where $\tilde{A}_0(\omega) = \tilde{A}_{m-1}(\omega)H_a(\omega) + \tilde{n}_m(\omega)$ and the appropriate kernel transforms $H_{1,m}(\omega)$ and $H_{3,m}(\omega)$ have to be used. It becomes immediately obvious that numerous mixing products between signal and noise arise from the evaluation of the product inside the convolution integral. These mixing products account for nonlinear interaction between signal and noise during propagation. However, a thorough analysis incorporating nonlinear signal-noise interactions yields complex expressions as the ones found in [109], which are not suitable as a basis for simple design rules. In order to simplify the analysis considerably, amplifier noise and consequently any

---
[3]The cumulated gain is defined such that $P(z) = P_0 \exp(G(z))$.

3 Single-Span Model

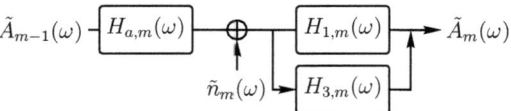

Figure 3.5: Block diagram of a fibre section with kernel transforms $H_{1,m}(\omega)$ and $H_{3,m}(\omega)$ preceded by an optical amplifier with transfer function $H_{a,m}(\omega)$ adding noise $\tilde{n}_m(\omega)$ to the signal.

nonlinear interaction between signal and noise is not considered in the following derivation. The transfer function $H_a(\omega)$ of the amplifiers is incorporated into the gain profile $G(z)$ as defined in Eq. (3.16). Therefore, the following results do not hold for systems where the nonlinear interaction between signal and noise is the dominant source of signal degradation.

Under the assumption of noiseless optical amplifiers, the field envelope at the output of the mth fibre section can be expressed with the recursion [analogous to Eq. (3.9)]

$$\tilde{A}_m(\omega) \approx H_{1,m}(\omega, L_m)\left(\tilde{A}_{m-1}(\omega) + j\iint_{-\infty}^{\infty} \eta_m(\Delta\Omega, L_m)\tilde{S}_{m-1}\,d\omega_1\,d\omega_2\right), \quad (3.18)$$

where A_{m-1} is the field envelope at the input of the mth fibre section, $\tilde{S}_i = \tilde{A}_i(\omega_1)\tilde{A}_i^*(\omega_2)$ and $H_{1,m}$, η_m and L_m are the first-order Volterra kernel transform, the NLTF and the length of the mth fibre section, respectively. Please note that since the nonlinear perturbation generated in each span is assumed to be very small, it is neglected when calculating $\tilde{S}_{m-1}$. With Eq. (3.18) the output field envelope $\tilde{A}_M(\omega)$ after transmission over M fibre sections is

$$\tilde{A}_M(\omega) \approx \left(\prod_{m=1}^{M} H_{1,m}(\omega, L_m)\right)\left(\tilde{A}_0(\omega) + \sum_{m=1}^{M}\delta_{NL,m}(\omega)\right), \quad (3.19)$$

where the nonlinear perturbation generated in the mth fibre section is

$$\delta_{NL,m}(\omega) = j\gamma_m\iint_{-\infty}^{\infty}\int_{z_{m-1}}^{z_m} e^{G(z)+jC(z)\Delta\Omega}\,dz\,\tilde{S}_0\,d\omega_1\,d\omega_2, \quad (3.20)$$

with $z_0 = 0$, $L_m = z_m - z_{m-1}$ and γ_m the length and nonlinear coefficient of the mth fibre and gain and dispersion profiles $G(z)$ and $C(z)$ as defined in Eq. (3.16)

3.2 Nonlinear Transfer Function

and (3.17). To get a similar notation as in the single-fibre case, the overall nonlinear perturbation can be noted as

$$\delta_{NL}(\omega) = \sum_{m=1}^{M} \delta_{NL,m}(\omega)$$
$$= j \iint_{-\infty}^{\infty} \eta(\Delta\Omega)\, \tilde{S}_0 \, d\omega_1 \, d\omega_2, \quad (3.21)$$

with the overall NLTF of a concatenation of M fibre sections

$$\eta(\Delta\Omega) = \sum_{m=1}^{M} \gamma_m \int_{z_{m-1}}^{z_m} e^{G(z) + jC(z)\Delta\Omega} \, dz. \quad (3.22)$$

It is easily verified that Eq. (3.11) is a special case of the above equation with $M = 1$, no dispersion precompensation and no distributed amplification. Further simplification is possible by noting that the product of first-order kernel transforms in Eq. (3.19) describes the residual gain $G(z_M)$ and dispersion $C(z_M)$ at the receiver. The overall first-order kernel transform of the concatenated fibres can thus be defined as

$$H_1(\omega) = \prod_{m=1}^{M} H_{1,m}(\omega, L_m) = e^{G(z_M) + jC(z_M)\Delta\Omega}. \quad (3.23)$$

Analogous to the solution for a single fibre in Eq. (3.9), Eq. (3.19) can now be written as

$$\tilde{A}_M(\omega) \approx H_1(\omega) \left(\tilde{A}_0(\omega) + \delta_{NL}(\omega) \right). \quad (3.24)$$

The key assumptions for Eq. (3.20), (3.22) and (3.24) to be valid are:

- Operation in a weakly nonlinear transmission regime and thus small nonlinear perturbation verifying $\int |\delta_{NL}(\omega)|^2 \, d\omega \ll \int |\tilde{A}_0(\omega)|^2 \, d\omega$ is assumed [108, 112].

- Volterra kernels of higher order than three are assumed to be negligible. Furthermore, the influence of nonlinear distortions on the nonlinear perturbation generated in subsequent spans is neglected. This condition may not be fulfilled for a large number of concatenated fibres or high input power.

3 Single-Span Model

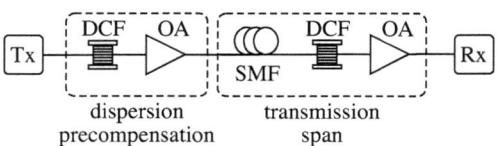

Figure 3.6: Schematic of a single-span system with a dispersion precompensation stage after the transmitter.

- Nonlinear interaction between amplifier noise and signal (e. g. Gordon-Mollenauer phase noise [119]) must be negligible. This condition may not be fulfilled for a large number of concatenated fibres or systems operating at low OSNR and can be problematic when dealing with phase-modulated signals.

3.2.3 Nonlinear Transfer Function of a Single-Span System

Eq. (3.22) describes the NLTF for the general case of arbitrary fibre-optic transmission lines. In the following, more confined, albeit common, scenarios are discussed. Fibre-optic transmission lines are usually organised in spans instead of fibre sections as basic building blocks. Each span comprises more than one fibre section and typically consists of a transmission fibre, inline dispersion compensation (e. g. a DCF) and means of signal amplification (e. g. an EDFA). Commonly, a certain amount of dispersion precompensation is used to predistort the launched signal at the transmitter and thereby reduce the impact of fibre nonlinearity on the signal [29, 120]. The most basic form of such a precompensated transmission line consists of just a single span and is depicted in Fig. 3.6.

For the following analysis, the input power into the DCFs is assumed small enough to treat them as linear dispersion-compensation devices and neglect their nonlinear impact. This means that they do not have to be considered as individual fibre sections and can be approximated as ideal lumped dispersion-compensation devices [cp. Fig. 3.4(c)]. Under this assumption, dispersion precompensation can be included in the NLTF simply through the cumulated dispersion profile $C(z)$. For the single-span system shown in Fig. 3.6, the cumulated dispersion can thus be written with Eq. (3.17) as $C(z) = C_{pre} + \beta_2 z$, where C_{pre} is the amount of dispersion precompensation in units ps^2 and β_2 is the GVD of the SMF. The optical amplifiers completely compensate for the fibre loss of preceding fibres. With Eq. (3.13) and (3.22) (with

3.2 Nonlinear Transfer Function

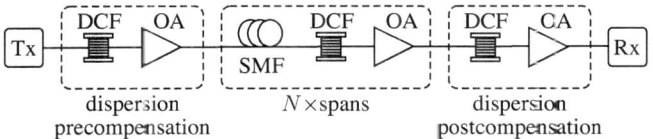

Figure 3.7: Schematic of a fibre-optic transmission line consisting of N identical spans and dispersion pre- and postcompensation stages.

$M = 1$) the transfer function of a single span with dispersion precompensation is

$$\eta(\Delta\Omega) \approx \eta_s(\Delta\Omega) e^{jC_{pre}\Delta\Omega}, \qquad (3.25)$$

where η_s is the NLTF of a single fibre according to Eq. (3.13). Introducing a precompensation thus only affects the phase of the NLTF and has no effect on its magnitude.

3.2.4 Nonlinear Transfer Function of Systems with Single-Periodic Dispersion Maps

A common scheme for dispersion management is the single-periodic dispersion map [29]. Ideally, the transmission line consists of N identical spans[4], each comprising of a transmission fibre, a DCF and an optical amplifier. A schematic of such a transmission line is shown in Fig. 3.7. In a transmission line consisting of multiple spans, the amount of dispersion compensation performed in each span adds another degree of freedom to the optimisation of a dispersion map. Usually, the dispersion is not fully compensated per span, leading to a certain amount of residual dispersion per span (RDPS). Assuming low input power into the DCF and thus negligible nonlinear impact, they act as ideal lumped dispersion compensation modules at positions z_m. In this case, they do not need to be treated as individual fibre sections and only affect the cumulated dispersion $C(z)$[5]. The cumulated dispersion $C(z)$ of a transmission line with dispersion precompensation and RDPS is shown in Fig. 3.8 (black line). Under the assumption of identical spans, length, fibre loss, GVD and nonlinear coefficients of the transmission fibres are the same in each span and Eq. (3.22) simplifies

---
[4]The number of spans is denoted by N in order to distinguish between spans and fibre sections.
[5]Due to this simplification, only transmission fibres are treated as fibre sections and it follows that $N = M$ for this special case.

3 Single-Span Model

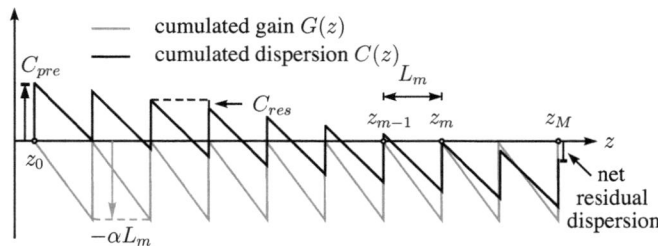

Figure 3.8: Profile of the cumulated gain $G(z)$ and cumulated dispersion $C(z)$ in a single-periodic dispersion map with dispersion precompensation C_{pre} and RDPS C_{res}.

to [114]

$$\eta(\Delta\Omega) = \frac{\gamma}{\alpha} \frac{1 - e^{-\alpha L + j\beta_2 L \Delta\Omega}}{1 - j\frac{\beta_2}{\alpha}\Delta\Omega} \sum_{n=1}^{N} e^{G_n + jC_n \Delta\Omega}, \qquad (3.26)$$

where L is the length of the transmission fibre, N is the number of identical spans and G_n and C_n are the cumulated gain and dispersion at the beginning of the nth span. Further simplification is possible by noting that loss is usually compensated per span, leading to $G_n = 0$. The cumulated gain for full reamplification per span is also shown in Fig. 3.8 (grey line). Assuming identical spans means that there is a constant RDPS C_{res} and the cumulated dispersion at the beginning of the nth span is $C_n = C_{pre} + (n-1)C_{res}$. In most cases, fibres are long with respect to their effective length such that $L \gg 1/\alpha$ [6]. In this case, the term in front of the sum in Eq. (3.26) is approximately equal to the NLTF η_s of a single fibre according to Eq. (3.13). Under these conditions, Eq. (3.26) can be written as

$$\eta(\Delta\Omega) \approx \eta_s(\Delta\Omega) e^{jC_{pre}\Delta\Omega} \sum_{n=1}^{N} e^{j(n-1)C_{res}\Delta\Omega}. \qquad (3.27)$$

---

[6] As long as $L \gg L_{eff}$ for all fibres, the length of fibres may actually vary and they need not be identical in this respect. Since the nonlinear phase shift is mainly generated within the effective length, length variations for $L \gg L_{eff}$ have no significant impact on the nonlinear perturbation. For example, varying fibre lengths between 50 km and 110 km for transmission of 5×10 Gb/s NRZ-OOK over 1200 km with 50 GHz channel spacing and launch power $P_0 = 3$ mW showed no significant difference to transmission over fibres with a uniform length of 80 km [121].

3.2 Nonlinear Transfer Function

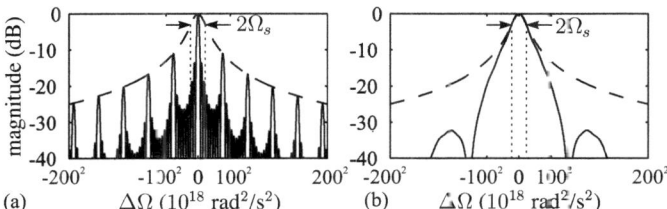

Figure 3.9: Solid lines show the magnitude of the nonlinear transfer function according to Eq. (3.28). The transmission line consists of 10×80 km SSMF spans with RDPS corresponding to (a) 40 km and (b) 2 km of uncompensated SSMF. Dashed lines indicate the magnitude of $\eta_s(\Delta\Omega)$ according to Eq. (3.13). Magnitudes are normalised to their respective maximum values.

Evaluating the geometric series in Eq. (3.27) yields

$$\eta(\Delta\Omega) \approx \eta_s(\Delta\Omega) \frac{e^{jNC_{res}\Delta\Omega}-1}{e^{jC_{res}\Delta\Omega}-1} e^{jC_{pre}\Delta\Omega}$$
$$= \eta_s(\Delta\Omega) \frac{\sin\left(\frac{NC_{res}}{2}\Delta\Omega\right)}{\sin\left(\frac{C_{res}}{2}\Delta\Omega\right)} e^{jC'_{pre}\Delta\Omega} \quad (3.28)$$

with

$$C'_{pre} = C_{pre} + \frac{N-1}{2}C_{res}. \quad (3.29)$$

In contrast to dispersion precompensation, introduction of RDPS thus has an influence on both phase and magnitude of the NLTF.

Fig. 3.9 shows the influence of RDPS on the magnitude of the NLTF for a transmission line consisting of $N = 10$ spans. The transmission fibre is a SSMF with fibre loss $\alpha = 0.2$ dB/km and dispersion parameter $D = 16$ ps/(nm·km). Dashed lines represent the magnitude of $\eta_s(\Delta\Omega)$ normalised to its maximum value. This corresponds to FWM efficiency for transmission over a single fibre [cp. Eq. (2.34)]. Furthermore, it also describes FWM efficiency for a transmission line with full in-line dispersion compensation per span, i.e. for $C_{res} = 0$ in Eq. (3.28). In contrast to that, solid lines show the NLTF of the transmission line with RDPS. Introduction of a RDPS makes the magnitude of the NLTF oscillate. The oscillation frequency depends on the amount of RDPS and the number of spans N, while the envelope of the oscillation is described by the NLTF $\eta_s(\Delta\Omega)$ of a single fibre. Again, the NLTF normalised to its maximum value corresponds to the FWM efficiency in systems with RDPS. For example, evaluating the magnitude of the NLTF for the special case

3 Single-Span Model

of a transmission line without dispersion compensation[7] corresponds to the FWM efficiency derived for such systems in [122]. Measurements of FWM efficiency performed in a dispersion-managed system [123] show very good agreement with the analytical description presented in [122]. A proper choice of RDPS can thus be used to reduce FWM efficiency in multi-span transmission lines.

In its present form, the NLTF for transmission lines with RDPS is very different from the NLTF $\eta_s(\Delta\Omega)$ for a single-span. However, for very small arguments of the sine-functions, i.e. for

$$\frac{N}{2}|C_{res}\Delta\Omega| \ll 1, \tag{3.30}$$

Eq. (3.28) can be approximated as

$$\eta(\Delta\Omega) \approx N\eta_s(\Delta\Omega) \exp\left(jC'_{pre}\Delta\Omega\right). \tag{3.31}$$

While the phase term is preserved, the magnitude is approximated by its envelope. Essentially, this is the N-fold transfer function of a single fibre with an additional phase shift governed by C'_{pre}. In this case, C'_{pre} can be interpreted as the amount of dispersion precompensation of an equivalent single-span system, which becomes clear when comparing Eq. (3.31) to Eq. (3.25). Within the validity of the first-order perturbation approach and Eq. (3.30), the NLTF of systems with arbitrary single-periodic dispersion maps can thus be described by the NLTF of simple single-span systems. However, due to the factor N in Eq. (3.31) the nonlinear perturbation generated over N spans is larger than that generated over a single span. This discrepancy can be approached by considering propagation of the complex envelope of the electric field normalised to the average launch power P_0 as

$$\tilde{u}(\omega, z) = \frac{\tilde{A}(\omega, z)}{\sqrt{P_0}}. \tag{3.32}$$

Substituting $\tilde{u}$ in Eq. (3.24) yields the normalised envelope after transmission over M fibre sections

$$\tilde{u}_M(\omega) = H_1(\omega) \left(\tilde{u}_0(\omega) + j \iint_{-\infty}^{\infty} \eta(\Delta\Omega) P_0 \right.$$
$$\left. \times \tilde{u}_0(\omega_1)\tilde{u}_0(\omega_2)\tilde{u}_0(\omega_3) \, d\omega_1 \, d\omega_2 \right). \tag{3.33}$$

---
[7]That means $C_{res} = \beta_2 L$ in Eq. (3.28).

3.2 Nonlinear Transfer Function

From this equation it becomes clear that the nonlinear perturbation depends on the product of NLTF and average launch power. For the case of single-periodic dispersion maps, this product can be written with Eq. (3.13) and (3.31) as

$$\eta(\Delta\Omega)P_0 \approx \frac{\Phi_{NL}}{1 - j\,\mathrm{sgn}\,(\beta_2)\frac{\Delta\Omega}{\Omega_s}} \exp\left(jC'_{pre}\Delta\Omega\right), \tag{3.34}$$

where

$$\Phi_{NL} = N\frac{\gamma P_0}{\alpha} = N\frac{L_{eff}}{L_{NL}} \tag{3.35}$$

is the average nonlinear phase shift. The generated nonlinear perturbations in the equivalent single-span system and N-span system will thus be comparable, provided that the average launch power P_0 is scaled such that the average nonlinear phase shift remains constant.

Unfortunately, the validity of the equivalent single-span approximation is very strictly limited by Eq. (3.30). Should the approximation Eq. (3.31) be at least accurate for $|\Delta\Omega| \leq \Omega_s$, it follows from Eq. (3.30) that $N|C_{res}| \ll 2|\beta_2|/\alpha$. This means that the cumulated residual dispersion over all spans has to be much less than the cumulated dispersion over twice the effective length of the transmission fibre. Extensive numerical simulations for 40 Gb/s OOK as well as DPSK have shown that the match in terms of required optical signal-to-noise ratio (ROSNR) penalty is reasonably good for

$$N|C_{res}| \leq 0.8\frac{|\beta_2|}{\alpha}. \tag{3.36}$$

For example, considering 5 spans of SSMF [with $D = 16$ ps/(nm·km)], $|D_{res}| \leq 60$ ps/nm [124, 125].

The effect of the approximation of sine functions in Eq. (3.28) is clarified in Fig. 3.9 Fig. 3.9(a) shows the magnitude of the NLTF for a system with large RDPS corresponding to 40 km uncompensated SSMF length. Clearly, the approximation of the main lobe, even for $|\Delta\Omega| < \Omega_s$ is very poor. On the other hand, the approximation is quite good for small RDPS as seen in Fig. 3.9(b), where the error is small for $|\Delta\Omega| \leq (100 \cdot 10^9 \,\frac{\mathrm{rad}}{\mathrm{s}})^2$.

39

3 Single-Span Model

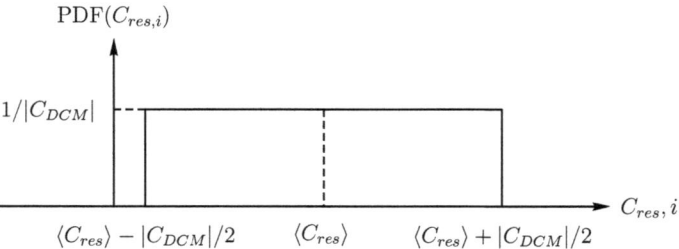

Figure 3.10: Probability density function (PDF) of the residual dispersion of an individual span. The uniform distribution has mean $\langle C_{res}\rangle$ and its boundaries are defined by the cumulated dispersion C_{DCM} of the employed DCM.

3.2.5 Nonlinear Transfer Function of Systems with Randomly Varying Residual Dispersion per Span

In practical transmission systems, the RDPS usually varies from span to span due to a certain granularity of commercially available dispersion-compensating modules (DCMs), e.g. DCM-20 which compensate 20 km of SSMF are commonly used in 10 Gb/s systems. Although one tries to approximate a certain nominal value of RDPS as closely as possible, the accuracy of this approximation is limited by the granularity of the available DCMs. Due to these practical reasons, the residual dispersion of an individual span can be viewed as a random number, drawn from a uniform distribution, whose mean value $\langle C_{res}\rangle$ is the nominal value of RDPS. The width of the uniform distribution equals the cumulated dispersion C_{DCM} of the employed DCMs. Fig. 3.10 shows the probability density function of the residual dispersion of an individual span $C_{res,i}$.

For the derivation of the NLTF of a transmission line with such a random RDPS, key assumptions from the last section pertaining to a transmission line with identical spans and constant RDPS still hold.

- The propagation in DCFs can be assumed as linear, such that they do not need to be considered as individual fibre sections.

- All transmission fibres are identical and are long compared to their effective length (i.e. $L \gg L_{eff}$).

3.2 Nonlinear Transfer Function

- Fibre loss is fully compensated per span, such that the cumulated gain at the input of the nth span is $G_n = 0$.

With these assumptions the general NLTF (3.22) simplifies to Eq. (3.26). Since $G_n = 0$, the NLTF for transmission lines with randomly varying RDPS can thus be derived by describing the cumulated dispersion C_n.

Let $\Delta C_{res,i}$ be the deviation of the RDPS of the ith span from the nominal RDPS $\langle C_{res} \rangle$, such that $C_{res,i} = \langle C_{res} \rangle + \Delta C_{res,i}$. The $\Delta C_{res,i}$ are independent and identically distributed random variables. They are uniformly distributed between $-|C_{DCM}|/2$ and $|C_{DCM}|/2$. The cumulated dispersion at the beginning of the nth span can then be expressed as

$$C_n = C_{pre} + (n-1)\langle C_{res} \rangle + \sum_{i=0}^{n-1} \Delta C_{res\, i}, \tag{3.37}$$

where $\Delta C_{res,0}$ accounts for a possible deviation of the precompensation from its nominal value[8]. The RDPS mismatch $\Delta C_{res,i}$ is uniformly distributed with zero mean and distribution boundaries defined by the cumulated dispersion C_{DCM} of the employed DCM. Taking above example of DCM-20, the maximum mismatch per span corresponds to 10 km of SSMF (i.e. $C_{DCM} = 400$ ps^2 and -200 ps$^2 \leq \Delta C_{res,i} \leq 200$ ps^2).

Assuming small deviations $|\Delta C_{res,i}| \ll |\beta_2/\alpha|$ for all spans and validity of Eq. (3.30) for $C_{res} = \langle C_{res} \rangle$ it can be shown that Eq. (3.26) may be approximated as

$$\eta(\Delta\Omega) \approx N\eta_s(\Delta\Omega)\, e^{j\frac{1}{N}\sum_{n=1}^{N} C_n \Delta\Omega}. \tag{3.38}$$

This leads to the same single-span representation as in Eq. (3.31) but with an equivalent precompensation of

$$C'_{pre} = \frac{1}{N}\sum_{n=1}^{N} C_n. \tag{3.39}$$

For the special case of periodic dispersion maps with non-random RDPS, above equation reduces to the expression given in Eq. (3.29).

---

[8] Please note that the nominal RDPS $\langle C_{res} \rangle$ is the mean of $\text{PDF}(C_{res,i})$. Therefore it is not necessarily the mean RDPS of a transmission line and it is only for $N \to \infty$ that $\lim \sum_{i=0}^{N-1} \Delta C_{res,i} = 0$.

3 Single-Span Model

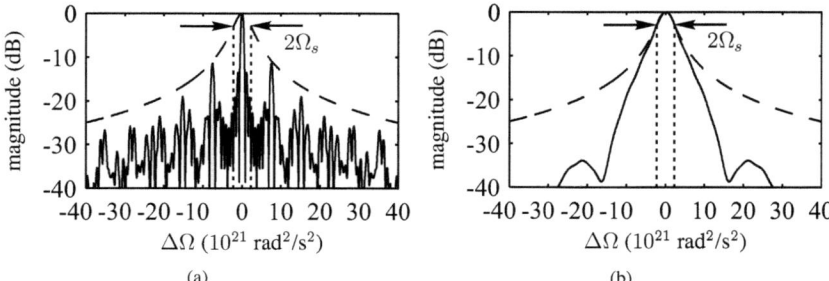

Figure 3.11: Magnitude of the nonlinear transfer function for 10×80 km SSMF and RDPS distributed with (a) $\langle C_{res} \rangle = -816$ ps^2, $C_{DCM} = 413$ ps^2 (corresponding to DCM-20) and (b) $\langle C_{res} \rangle = -41$ ps^2, $C_{DCM} = 41$ ps^2 (corresponding to DCM-2). Dashed lines indicate the magnitude of $\eta_s(\Delta\Omega)$.

Fig. 3.11 shows the magnitude of the NLTF for the same system configurations as in Fig. 3.9 but with random variations of RDPS. Randomisation of the RDPS destroys the regular resonances in the NLTF's magnitude. In particular, this effect can be observed for large variations of RDPS [cp. Figs. 3.9(a) and 3.11(a)]. To verify applicability of the equivalent single-span model and quantify the impact of random variations of RDPS, extensive numerical simulations were conducted. The results are discussed in chapter 4, section 4.2.2.

3.3 Scaling Laws

In the preceding section, it was shown that the first-order nonlinear perturbation of a signal at the end of a transmission line consisting of N spans can be approximated by the nonlinear perturbation generated in a single span. This single-span approximation is in agreement with important design rules derived in recent years, pertaining to the dependence of nonlinear impairments on cumulated nonlinear phase [126, 127], bit rate and fibre dispersion [128, 129] as well as dispersion map [59, 130]. These design rules define the parameters of the equivalent single-span system and thus the approximate first-order nonlinear perturbation. The three parameters defining the single-span system are the 3-dB bandwidth Ω_s of the NLTF, the equivalent precompensation C'_{pre} and the cumulated nonlinear phase shift Φ_{NL} according to Eq. (3.35). In many studies, all of these parameters have been found to significantly influence

3.3 Scaling Laws

the performance of fibre-optic transmission systems. In the following, each of these three parameters is discussed in greater detail and put into context with existing research.

3.3.1 3-dB Bandwidth of the Nonlinear Transfer Function

The 3-dB bandwidth of the single-span NLTF [Eq. (3.14)] depends entirely on the employed transmission fibre, more precisely on its attenuation coefficient and GVD. It essentially describes the 3-dB bandwidth of FWM efficiency.

In order to highlight the significance of Ω_s for propagation of the electric field in a single-mode fibre, the NLSE can be written in a parameterised form. For the derivation of this form it is useful to introduce a normalised amplitude function independent of fibre loss as follows

$$U(z,T) = \frac{A(z,T)}{\sqrt{P_0}\exp\left(-\frac{\alpha}{2}z\right)}. \tag{3.40}$$

Substituting $U(z,T)$ into the NLSE (2.17) yields

$$\frac{\partial U(z,T)}{\partial z} = -j\frac{\beta_2}{2}\frac{\partial^2 U(z,T)}{\partial T^2} + j\gamma P_0 \exp(-\alpha z)\,|U(z,T)|^2 U(z,T). \tag{3.41}$$

Writing this equation with dimensionless variables

$$\zeta = \alpha z \tag{3.42}$$

and

$$\tau = TR_s \tag{3.43}$$

leads to

$$\frac{\partial U(\zeta,\tau)}{\partial \zeta} = -j\mathrm{sgn}\,(\beta_2)\frac{C_1}{2}\frac{\partial^2 U(\zeta,\tau)}{\partial \tau^2} + j\Phi_{NL}\exp(-\zeta)\,|U(\zeta,\tau)|^2 U(\zeta,\tau), \tag{3.44}$$

where Φ_{NL} is the nonlinear phase shift according to Eq. (3.35) (with $N=1$) and C_1 is a dimensionless parameter defined as

$$C_1 = \frac{R_s^2}{\Omega_s} = R_s^2\frac{|\beta_2|}{\alpha}. \tag{3.45}$$

43

3 Single-Span Model

The parameter C_1 can be interpreted as a normalisation of the cumulated dispersion of the effective length by the symbol rate. In the following, it is therefore referred to as normalised cumulated dispersion. It is clearly seen from Eq. (3.44) that normalised dispersion C_1 and nonlinear phase shift Φ_{NL} govern the evolution of the electric field in the time domain.

The normalised cumulated dispersion C_1 has been shown to be a direct measure for the maximum number of overlapping and nonlinearly interacting pulses in a single wavelength channel within the effective length of a fibre [46]. Furthermore, it was demonstrated by numerical simulations in [129] and without inclusion of fibre loss in [128, 131, 132] that it universally describes the impact of intrachannel nonlinearities. Interestingly, this result has both a frequency-domain and a length-scale interpretation since for long fibres (with $L \gg L_{eff}$) $R_s^2/\Omega_s = L_{eff}/L_D$.

The frequency-domain NLSE can be parameterised in a similar fashion. Substituting the Fourier transform $\tilde{U}(z,\omega)$ of $U(z,T)$ defined in Eq. (3.40) into the NLSE in the frequency domain (3.3) yields

$$\frac{\partial \tilde{U}(z,\omega)}{\partial z} = j\omega^2 \frac{\beta_2}{2} \tilde{U}(z,\omega) + j\gamma P_0 \exp(-\alpha z) \\ \times \iint_{-\infty}^{\infty} \tilde{U}(z,\omega_1)\tilde{U}^*(z,\omega_2)\tilde{U}(z,\omega_3)\,\mathrm{d}\omega_1\,\mathrm{d}\omega_2. \tag{3.46}$$

Writing Eq. (3.46) with dimensionless variables ζ as in Eq. (3.42) and

$$\Omega = \frac{\omega}{2\pi \Delta f_{ch}}, \tag{3.47}$$

leads with Eq. (3.35) ($N=1$) to

$$\frac{\partial \tilde{U}(\zeta,\Omega)}{\partial \zeta} = 2j\pi^2 \Omega^2 \mathrm{sgn}(\beta_2) C_2 \tilde{U}(\zeta,\Omega) + j\Phi_{NL}\exp(-\zeta) \\ \times \iint_{-\infty}^{\infty} \tilde{U}(\zeta,\omega_1)\tilde{U}^*(\zeta,\omega_2)\tilde{U}(\zeta,\Omega,\omega_1,\omega_2)\,\mathrm{d}\omega_1\,\mathrm{d}\omega_2, \tag{3.48}$$

where

$$C_2 = \frac{\Delta f_{ch}^2}{\Omega_s} = \Delta f_{ch}^2 \frac{|\beta_2|}{\alpha}. \tag{3.49}$$

Similar to C_1, the dimensionless parameter C_2 describes the dispersion cumulated over the effective length but in this case normalised by the WDM channel spacing

3.3 Scaling Laws

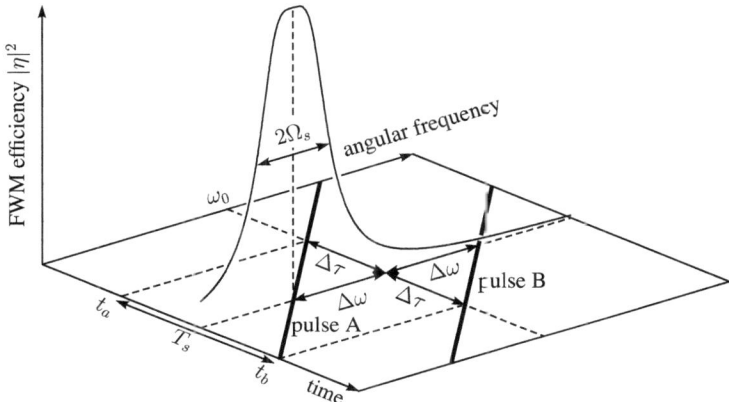

Figure 3.12: Spectrogram representation of two dispersive Dirac pulses (bold lines) and the formation of IFWM. Pulses A and B are centred at time instants t_a and t_b and at reference angular frequency ω_0 in time and frequency domain, respectively. Their separation in the frequency domain is $2\Delta\omega$. The vertical axis represents the efficiency of the FWM process with 3-dB bandwidth Ω_s.

Δf_{ch}. Eq. (3.48) demonstrates that the frequency-domain NLSE can be parameterised by normalised dispersion C_2 and nonlinear phase shift Φ_{NL}. For constant spectral efficiency, parameters C_1 and C_2 are equivalent. However, this equivalence is destroyed when varying spectral efficiency is considered. This will become clear in the following, where scaling of FWM and IFWM efficiency with C_1 and C_2 is analysed.

Now that the two parameters C_1 and C_2 are established, the question arises, how do interchannel and intrachannel nonlinearities scale with them? In the following scaling of these different nonlinear effects is discussed using the example of FWM efficiency. First, the efficiency of IFWM between two overlapping pulses is derived. Subsequently, scaling of IFWM and FWM efficiencies with parameters C_1 and C_2 are compared.

As explained in section 2.2.1, the effect of IFWM originates in FWM between red and blue frequency components of overlapping pulses. For a simple analysis of the scaling of IFWM efficiency with the normalised cumulated dispersion C_1, the case of two dispersive Dirac pulses is considered. They are defined at $z = 0$ as $U_a(0,t) = \delta(t-t_a)$ and $U_b(0,t) = \delta(t-t_b)$, where $\delta(x)$ is Dirac's delta function. The respective Fourier transforms have magnitudes $|\tilde{U}_a(0,\omega)|^2 = 1$ and $|\tilde{U}_b(0,\omega)|^2 = 1$.

3 Single-Span Model

A spectrogram of two such pulses affected by dispersion is shown in Fig. 3.12 (bold lines in the time-frequency plane). In the time domain, the two pulses are centred at $t = t_a$ (pulse A) and $t = t_b$ (pulse B) such that the symbol duration is $T_s = t_b - t_a$. The group-delay difference $\Delta\tau$ between the optical carrier at reference angular frequency ω_0 and a spectral component spaced $\Delta\omega$ apart can be approximated by

$$\Delta\tau \approx \beta_2 \Delta\omega z. \tag{3.50}$$

For $z = 1/\alpha \approx L_{eff}$, the frequency difference between overlapping pulses A and B for $|\Delta\tau| = T_s/2$ is thus determined as

$$2\Delta\omega = \frac{\alpha}{|\beta_2| R_s}. \tag{3.51}$$

The efficiency of IFWM between the two pulses can then be calculated with Eq. (2.34) and $\Delta\Omega = (2\Delta\omega)^2$ to be

$$|\eta_{\mathrm{IFWM}}|^2 = \frac{1}{1 + \left(\frac{\alpha}{\beta_2 R_s^2}\right)^2} = \frac{1}{1 + \frac{1}{C_1^2}}. \tag{3.52}$$

It is immediately obvious that the efficiency of IFWM monotonously increases with C_1 and that $\lim_{C_1 \to 0} |\eta_{\mathrm{IFWM}}|^2 = 0$ and $\lim_{C_1 \to \infty} |\eta_{\mathrm{IFWM}}|^2 = 1$.

However, for small values of C_1, IFWM is not so much limited by FWM efficiency but rather by the lack of sufficient pulse overlap. In order to find a lower bound on C_1 it can be assumed that for practical modulated signals, only spectral components with $\Delta\omega \leq 2\pi R_s$ carry significant power and that the most significant nonlinear interactions occur within the effective length (i.e. $z \leq L_{eff}$). The condition for pulse overlap is $|\Delta\tau| \geq T_s/2$. With Eq. (3.50), $\Delta\omega = 2\pi R_s$ and $z = 1/\alpha \approx L_{eff}$ the condition for nonlinearly significant pulse overlap is

$$4\pi \frac{|\beta_2|}{\alpha} R_s^2 \geq 1, \text{ i.e. } C_1 \geq \frac{1}{4\pi}. \tag{3.53}$$

This implies that IXPM and IFWM do not occur for $C_1 < 1/(4\pi)$. For transmission over SSMF, this corresponds to a symbol rate of 13.5 GBd, which is in good agreement with the results in [47].

3.3 Scaling Laws

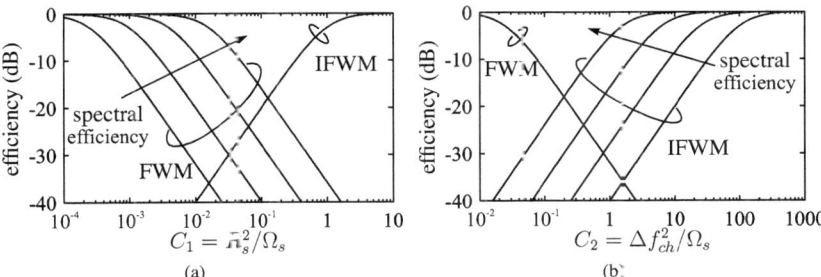

Figure 3.13: Efficiency of FWM and IFWM as functions of normalised cumulated dispersion (a) with respect to symbol rate R_s and (b) to channel spacing Δf_{ch}. The four shown cases of spectral efficiency are $0.1, 0.2, 0.4$ and 0.8 b/s/Hz (binary modulation assumed).

Similarly, the efficiency of FWM between adjacent WDM channels is determined by Eq. (2.34) (with $\Delta \Omega = 4\pi^2 \Delta f_{ch}^2$) as

$$|\eta_{\text{FWM}}|^2 = \frac{1}{1+\left(4\pi^2 \frac{\Delta f_{ch}^2 \beta_2}{\alpha}\right)^2} = \frac{1}{1+(4\pi^2 C_2)^2}. \quad (3.54)$$

The efficiency of FWM decreases monotonously with the normalised dispersion C_2 such that $\lim_{C_2 \to 0} |\eta_{\text{FWM}}|^2 = 1$ and $\lim_{C_2 \to \infty} |\eta_{\text{FWM}}|^2 = 0$.

The efficiency of FWM and IFWM is plotted in Fig. 3.13 as a function of normalised cumulated dispersion for spectral efficiencies $0.1, 0.2, 0.4$ and 0.8 b/s/Hz. When normalising the cumulated dispersion to the symbol rate, the channel spacing and thus FWM efficiency become a function of spectral efficiency [Fig. 3.13(a)]. Therefore this normalisation applies to systems where the major impairments come from intrachannel nonlinearities. On the other hand, normalising the cumulated dispersion to the WDM channel spacing makes symbol rate and thus efficiency of IFWM a function of spectral efficiency [Fig. 3.13(b)]. Due to this dualism it can be concluded that there is no single normalisation describing both FWM and IFWM for arbitrary spectral efficiency. Another implication of Fig. 3.13 is that interchannel nonlinearities dominate for small normalised dispersion, while intrachannel nonlinearities do so for large normalised dispersion. Optimum system configurations are likely located in-between these two extremes, where neither FWM nor IFWM reach their full efficiency. Furthermore, the minimum achievable efficiency depends on the spectral efficiency. As a consequence, increasing the spectral efficiency while keeping

3 Single-Span Model

the same modulation format will always result in a larger nonlinear penalty. These guesses will be verified by numerical simulations in the next chapter.

3.3.2 Equivalent Precompensation

The equivalent precompensation [Eq. (3.29) and (3.39)] describes the influence of the dispersion map and its parameters precompensation and RDPS. In the equivalent single-span model, the dispersion map only changes the phase of the NLTF, while its impact on the magnitude is neglected.

In the context of dispersion-map optimisation in transmission systems with a per-channel bit rate of 40 Gb/s, it was found that the optimum amount of precompensation depends on the RDPS in a linear fashion as described by the equivalent precompensation in Eq. (3.29). One approach to minimise the impact of intrachannel nonlinearities, which was proposed by Killey et al, is to minimise the pulse width within the effective length and thus minimise pulse overlap. It follows from this consideration that the optimum amount of dispersion precompensation is [59]

$$D_{pre} = -\frac{D}{\alpha}\ln\left(\frac{2}{1+\exp(-\alpha L)}\right) - \frac{N}{2}D_{res}. \qquad (3.55)$$

A very similar rule for optimum precompensation was found by means of numerical simulation and was verified for transmission over different fibre types. According to this rule, the optimum precompensation is [133]

$$D_{pre} = -\frac{N-1}{2}D_{res} + K, \qquad (3.56)$$

where K is the optimum value of dispersion precompensation for a system with full inline dispersion compensation. In later work, Eq. (3.56) was also derived analytically. This was done by applying small-signal analysis to determine the amplitude variance of the received signal induced by PM-to-AM conversion along the transmission line [130]. From this analytical derivation follows that $K \approx -D/\alpha$ is the optimum precompensation. This result is similar to Eq. (3.55) and is valid for $L \gg L_{eff}$ in systems limited by intrachannel nonlinearities. In the scope of the equivalent single-span model, the optimum K is determined by the optimum precompensation of the single-span system.

3.3.3 Nonlinear Phase Shift

The cumulated nonlinear phase shift [Eq. (3.35)] defines the general strength of the nonlinear perturbation. It depends on the power of the transmitted signal, as well as on the number of spans and the employed transmission fibre (nonlinear coefficient and attenuation coefficient).

The cumulated nonlinear phase shift has been shown to be strongly correlated to the expected penalty in many system configurations [126,127]. Furthermore, it can give an indication of the optimum net residual dispersion at the receiver [48].

4 Application to the Design of Fibre-Optic Transmission Systems

In this chapter, the theory derived in the previous chapter is verified by numerical simulations. First, section 4.1 describes the numerical modelling of fibre-optic communication systems, encompassing transmitters and receivers for various amplitude-shift keying (ASK) and phase-shift keying (PSK) modulation formats, the transmission line itself and last but not least criteria and methods to evaluate system performance. In section 4.2, the equivalent single-span model is verified for two exemplary configurations of practical interest. The first one represents 10 Gb/s-based legacy systems, the second one a possible 40 Gb/s scenario. While the first configuration allows verification of the equivalent single-span model for impairments caused by interchannel nonlinearities, the latter does so for intrachannel nonlinearities. Finally, section 4.3 presents applications of the model to the design of systems with multi-level modulation formats, varying spectral efficiency and electronic precompensation of intrachannel nonlinearities.

4.1 Simulation Model

This section briefly introduces the simulation environment and outlines necessary approximations due to trade-offs between simulation accuracy and available computation time. All simulations were performed with the commercially available software package VPItransmissionMaker^{TM}, complemented by Tcl/TK scripting as well as cosimulation programmed in Matlab$^{®}$. A block diagram for simulation of a generic WDM fibre-optic transmission system is depicted in Fig. 4.1. It consists of transmitters, a WDM multiplexer, dispersion precompensation stage, N fibre spans, a

4 Application to the Design of Fibre-Optic Transmission Systems

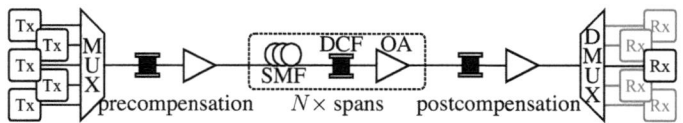

Figure 4.1: Generic setup of a WDM transmission system consisting of N spans.

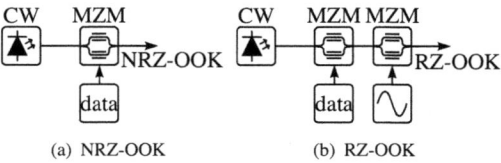

Figure 4.2: Schematic of basic OOK transmitters

postcompensation stage, WDM demultiplexer and receivers. All components are discussed consecutively, starting with transmitters in section 4.1.1, continuing with the transmission line in section 4.1.2 and receivers in section 4.1.3. Finally, section 4.1.4 presents a discussion about evaluation of signal quality at the receiver and system performance in general.

4.1.1 Transmitter

In optical communication systems information can be encoded into amplitude, phase, frequency or state of polarisation of the optical carrier. However, so far only ASK and DPSK have seen noteworthy deployment in commercial fibre-optic communication systems. This thesis concentrates on these modulation schemes, namely binary ASK, often referred to as on-off keying (OOK), differential binary phase-shift keying (DBPSK) and differential quadrature phase-shift keying (DQPSK). Extensive discussion of these modulation formats, their generation and reception as well as impact of linear and nonlinear impairments can be found e.g. in [30, 134, 135].

A schematic of NRZ-OOK and RZ-OOK transmitters is shown in Fig. 4.2. The continuous-wave (CW) lasers are considered as ideal monochromatic light sources, i.e. with zero linewidth. In systems with phase-modulated signals, phase noise of the laser source leads to additional penalties and places stringent limits on laser linewidth [136, 137]. These penalties are neglected here. The CW laser light is

4.1 Simulation Model

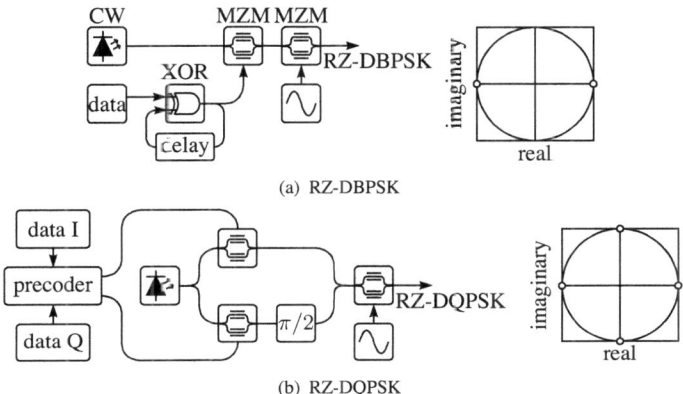

Figure 4.3: Schematics of (a) DBPSK and (b) DQPSK transmitters and associated constellation diagrams. [141]

intensity modulated with a Mach-Zehnder modulator (MZM) in push-pull operation, which is driven by the data signal. This mode of operation guarantees a chirp-free modulation [138]. In case of RZ-OOK, the data modulator is followed by a second sinusoidally driven MZM acting as a pulse carver. Depending on the bias point of the MZM as well as on frequency and amplitude of the sinusoidal drive signal, pulses with duty cycles of 33%, 50% and 67% can be generated [30]. In the following, only pulses with a duty cycle of 33% are considered. The extinction of the generated signals is assumed to be ideal.

It should be noted that some conclusions presented later in this section are subject to change based on the extinction ratio of the optical signals. For example, it has been reported in [139] that optimum dispersion maps for signals with very low extinction ratio ($ER \leq 10$ dB) differ from optimum maps derived in [59] for signals with larger extinction ratio. However, a signal extinction ratio of 10 dB is also the minimum recommended by the ITU [140]. Therefore, all presented results should be applicable to systems conforming with this ITU recommendation.

Schematics of RZ-DBPSK and RZ-DQPSK transmitters are shown in Fig. 4.3(a) and 4.3(b), respectively. The RZ-DBPSK transmitter is comparable in complexity to a RZ-OOK transmitter, both use one MZM for data modulation and one MZM for pulse carving. DPSK adds some complexity in the electronic domain since a

53

4 Application to the Design of Fibre-Optic Transmission Systems

precoder is necessary to differentially encode the data. For simulation purposes a logical XOR-gate with a delayed feedback-loop is used. The feedback signal is delayed by one symbol period. In practical realisations this is often replaced by an AND-gate followed by a T-flip-flop due to simple implementation of the same function [142, 143]. The MZM used for data modulation is biased at minimum transmission. This ensures a rectangular modulation of the phase of the electric field, where the data is encoded in a phase difference of either zero or π. The two resulting signal states are shown in the constellation diagram in Fig. 4.3(a).

The need for higher spectral efficiency and better tolerances towards linear impairments such as GVD has led to the development of optical multilevel modulation formats. A particularly successful modulation format is DQPSK, which uses a set of four possible phase differences two encode two bits per transmitted symbol. The transmitter and the resulting four signal states are shown in Fig. 4.3(b). Following [141] DQPSK is implemented here with two MZMs in a Mach-Zehnder superstructure. Unlike DBPSK this adds considerable complexity in the optical domain compared to RZ-OOK. Moreover, the electric precoder has to implement a more complex logical function in order to use direct detection [141]. A method to reduce the optical complexity of RZ-DQPSK generation by using a serial approach with two MZMs followed by a phase modulator has been proposed in [144].

For numerical simulations of high-speed optical transmission systems, a proper choice of the transmitted digital data is vital. In order to accurately model bit pattern dependent intrachannel nonlinear effects such as IFWM and IXPM, the length and content of the transmitted binary sequence has to be chosen adequately [46]. Pseudo-random binary sequences of length 2^m containing every possible permutation of m bits are particularly suited for this task. Such a sequence is commonly called a de Bruijn binary sequence (DBBS) after N. G. de Bruijn, who derived the number of existing DBBSs of a certain order [145]. The necessary length of the DBBS is still a subject of research and depends on the specific system configuration and target bit-error ratio (BER) [146, 147]. Unless stated otherwise, all simulations in this thesis use a DBBS of order ten, since that constitutes a good trade-off between accuracy and simulation time. Additional care has to be taken when multilevel modulation formats are considered. In case of DQPSK transmission, binary sequences do not lead to sufficiently stable results and have to be replaced by pseudo-random quaternary sequences (PRQSs) [148]. For all DQPSK simulations, a PRQS of length 4^5 is used. It is constructed by interleaving two DBBSs of length 2^{10} with an appropriate cyclic

4.1 Simulation Model

shift of ± 5 bits [149].

Besides these single channel considerations, the accuracy of numerical simulations of WDM systems has to be carefully weighed against needed computation time. The impact of interchannel nonlinear effects (FWM and XPM) depends on the bit patterns transmitted in interfering WDM channels and in case of FWM also on phase matching with these channels. In realistic systems, many WDM channels are transmitted simultaneously, such that the walk-off between channels spaced far away can add up to large numbers of bits. Furthermore, the phase relationship between WDM channels is usually random[1]. It is futile to attempt to simulate all possible combinations of bits in the observed channel and all interfering channels. A possible solution to this dilemma is to use a statistic approach. Similar to the simulation technique proposed in [151] for coherent WDM crosstalk, a system can be simulated numerous times with random timing and phase of each channel. From these simulation runs a distribution of the resulting BER is obtained. Since this procedure requires many simulation runs to be meaningful, it is very time consuming. Such a simulation procedure is only implemented for the analysis of DBPSK and DQPSK transmission in section 4.3.1 and implications on the results are discussed. Because of time constraints, one arbitrarily chosen constellation of timings and phases is used for other investigations.

The multiplexer filters are modelled as second order Gaussian bandpass filters with a 3-dB bandwidth of twice the symbol rate.

4.1.2 Transmission Line

A configuration of a fibre-optic transmission system with single-stage optical amplifiers is shown in Fig. 4.1. Propagation of the complex envelope of the electric field through the fibres is computed by the split-step Fourier method, which is implemented in the fibre model of VPItransmissionMaker^{TM}. The transmission line consists of a dispersion precompensation stage at the transmitter, a number of N identical transmission spans and a postcompensation stage before the receiver. Precompensation and postcompensation stages are modelled as purely linear dispersive elements without any nonlinear impact. Each of the N spans consists of a single-mode

---

[1] Although it has been demonstrated that the phase relationship between WDM channels can be controlled in point-to-point links [150] this technique is principally not applicable in networks with optical add-drop multiplexers.

4 Application to the Design of Fibre-Optic Transmission Systems

symbol rate (GBd)	fibre type		
	NZDSF	SSMF	SLAF
10	0.011	0.045	0.059
40	0.18	0.71	0.94
100	1.1	4.4	5.9
160	2.9	11	15

Table 4.1: Values of normalised dispersion $C_1 = R_s^2/\Omega_s$ for some combinations of symbol rate and fibre type. Values are rounded to the two most significant digits.

fibre (SMF), a dispersion-compensating fibre (DCF) and an optical amplifier (OA). The SMF is modelled with fibre loss $\alpha = 0.2$ dB/km, nonlinear index coefficient $n_2 = 2.6 \cdot 10^{-20}$ m²/W and effective core area $A_{eff} = 80$ µm² (corresponding to typical values for SSMF). At wavelength $\lambda = 1.55$ µm this results in a nonlinear coefficient $\gamma = 1.31$ W^{-1}km^{-1}. The GVD of the SMF varies. Second-order GVD (i.e. dispersion slope) and the nonlinear impact of inline DCFs have not been considered. The optical amplifier compensates the span loss and is modelled as noiseless, since the ASE-noise is treated analytically in the receiver model. The postcompensation stage in front of the receiver is adjusted such that the net residual dispersion is zero.

In order to easily identify practical system configurations in the normalised notation, tables 4.1 and 4.2 show exemplary values of normalised dispersion C_1 and C_2, respectively. There are mainly three common fibre types in present-day systems, NZDSF, SSMF and super-large-effective-area fibre (SLAF). Their fibre loss and dispersion parameters used for tables 4.1 and 4.2 are summarised in table 4.3. Although most simulations are performed with fibre loss and nonlinear coefficient typical for SSMF, the results can be scaled to fibres having different fibre loss and nonlinear coefficient by simply adjusting the launch power for constant nonlinear phase shift according to Eq. (3.35).

4.1.3 Receiver

Optical receivers are either based on direct detection or coherent detection of the incoming light. Direct detection receivers convert the optical intensity to an electric current. In this process all information about the optical phase is lost. Contrary

4.1 Simulation Model

channel spacing	fibre type		
(GHz)	NZDSF	SSMF	SLAF
25	0.071	0.28	0.37
50	0.28	1.1	1.5
100	1.1	4.4	6
200	4.6	18	23
400	18	71	94

Table 4.2: Values of normalised dispersion $C_2 = \Delta f_{ch}^2/\Omega_s$ for some combinations of channel spacing and fibre type. Values are rounded to the two most significant digits.

fibre type	fibre parameters		
	fibre loss α (dB/km)	dispersion parameter D (ps/(nm·km))	effective core area A_{eff} (µm^2)
NZDSF	0.22	4.5	72
SSMF	0.2	16	80
SLAF	0.19	20	120

Table 4.3: Parameters of fibre types used in tables 4.1 and 4.2.

to that, coherent detection, where the photodiode is used as a mixing device for the incoming optical signal and light from a local oscillator, preserves information about the optical field's amplitude and phase. This makes coherent systems particularly suited for electronic mitigation of transmission impairments at the receiver end [152,153]. Although only direct detection is considered here, qualitative results concerning the scaling of fibre nonlinearities should also be applicable to coherent systems, since the generated nonlinear perturbation does not depend on the type of receiver.

A schematic of a direct-detection receiver for ASK signals is shown in Fig. 4.4(a). The signal is noise-loaded with additive white Gaussian noise (AWGN) before the optical band-pass filter (BPF). The receiver consists of a photodiode (PD) followed by an electrical low-pass filter (LPF). After low-pass filtering and sampling, the digital data is recovered at the decision gate. Optical BPFs are modelled as second order Gaussian filters. This constitutes a good approximation for the transfer function of optical thin-film filters [154]. The electrical post-detection filter is a 5th order Bessel low-pass filter with a 3-dB cut-off frequency of $0.7 \times R_s$. This configuration of filter bandwidths is close to optimal for all considered modulation formats [155,156].

4 Application to the Design of Fibre-Optic Transmission Systems

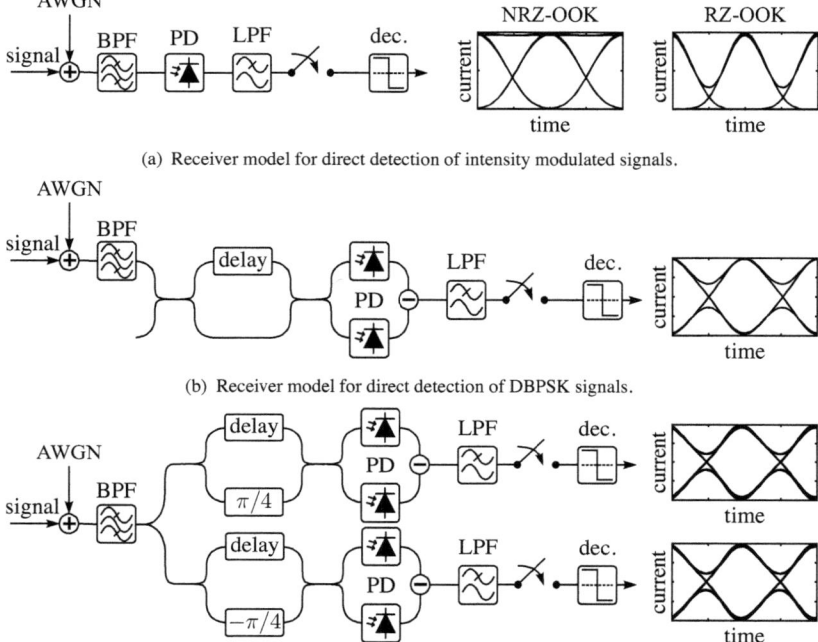

Figure 4.4: Schematics of (a) the receiver model for direct detection of intensity modulated signals, (b) receiver model for differential demodulation and balanced detection of DBPSK and (c) the same for DQPSK.

Back-to-back eye diagrams of NRZ-OOK and 33% RZ-OOK are shown on the right-hand side of Fig. 4.4(a).

The schematic of a direct detection receiver for DBPSK and the resulting back-to-back eye diagram are shown in Fig. 4.4(b). The delay interferometer with a delay of one symbol period demodulates the differential phase into intensity modulation at both output ports. The optical signal is then detected by two balanced photodiodes. In simulations conducted for this thesis, all components are assumed to work in ideal operating conditions. However, in practical systems these conditions can not be guaranteed due to e.g. manufacturing tolerances of the components. This

4.1 Simulation Model

leads to additional penalties [156–158] which are not considered in this thesis in order to remain as general as possible. A comprehensive analysis of transmitter and receiver structures for DBPSK and DQPSK which also covers aspects of non-ideal components can be found in [159].

The DQPSK receiver depicted in Fig. 4.4(c) follows the same principle as that for DBPSK. It consists of two DBPSK receivers, with $\pm\pi/4$ phase shifts in one arm of each delay interferometer [141]. Compared to DBPSK, this doubles the number of components in the optical domain as well as in the electrical domain. However, at the same bit rate DQPSK has only half the symbol rate, which lowers the required speed of electrical components. Also shown are back-to-back eye diagrams of both tributaries.

4.1.4 Evaluation of System Performance

The most important quality criterion for digital communication systems is the sustained BER, which is the probability of erroneously deciding the value of a received bit. The BER can be accurately estimated based on a Karhunen-Loève expansion of the ASE-noise before the optical demultiplexer filter and subsequent saddle-point approximation of the integral over the resulting moment generating function [160] This semi-analytic approach is embedded in a Matlab® cosimulation environment of VPItransmissionMaker^{TM} and was implemented by Sebastian Randel [154]. In this model, the noise is treated analytically and assumed to be additive white Gaussian noise (AWGN) before the demultiplexer filter, which is the case in many practical system configurations. With only minor adjustments the model is also suited for estimation of the BER in DBPSK and DQPSK systems with interferometric demodulation and direct detection [137, 161]. However, the assumption of noise loading at the receiver implies that nonlinear interactions between signal and noise along the transmission line (such as Gordon-Mollenauer phase noise [119]) are not considered. Since nonlinear phase noise is problematic mainly for PSK, an estimation of the impact of nonlinear phase noise is given in the context of DPSK transmission discussed in section 4.3.1.

A digital communication system is usually designed to guarantee operation below a certain target BER. In fibre-optic communication systems, the BER depends on

4 Application to the Design of Fibre-Optic Transmission Systems

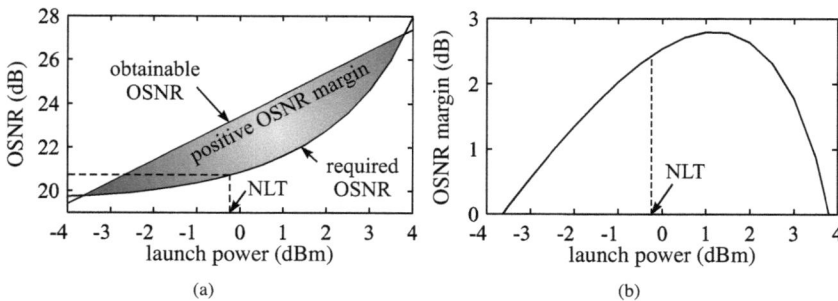

Figure 4.5: (a) Obtained and required OSNR as a function of launch power for transmission of 40 Gb/s NRZ-OOK over 10×80 km SSMF. The nonlinear threshold (NLT) is defined as the launch power resulting in 1 dB ROSNR penalty. (b) OSNR margin as a function of launch power. The NLT does not coincide with the launch power giving maximum OSNR margin.

the OSNR at the receiver. Time-invariance of the system assumed, a required optical signal-to-noise ratio (ROSNR) can be determined for operation below a certain BER. Unless stated otherwise all results are based on the ROSNR for BER = 10^{-9}. The ROSNR for the considered modulation formats in back-to-back configuration is summarised in Tab. 4.4. The actual OSNR at the receiver depends on the system configuration as in Eq. (2.3). The difference between ROSNR and actual OSNR is called OSNR margin and is used to safeguard against time-dependent impairments such as component aging, drift of system parameters with temperature and so forth. In a fixed transmission line, the only way to increase the OSNR at the receiver is to increase the launch power at the transmitter. Unfortunately, nonlinear impairments become stronger with increased launch power. Above a certain power threshold, the penalty in ROSNR due to nonlinear impairments outpaces the gain in OSNR at the receiver and a further increase of launch power becomes pointless. This principle is illustrated in Fig. 4.5(a), where ROSNR and obtained OSNR are plotted as functions of launch power for transmission of 40 Gb/s NRZ-OOK over 10×80 km SSMF. The noise figure of the optical amplifiers is assumed to be 5 dB, which is quite conservative, given that noise figures of about 4 dB are used for transoceanic fibre-optic transmission lines [162, 163] and that a noise figure of slightly above 3 dB should in principle be possible to obtain [164]. The resulting OSNR margin of the system is plotted in Fig. 4.5(b). In order to obtain a good estimate on where the maximum OSNR margin is reached, it is practical to define the nonlinear threshold (NLT) as the launch power leading to a 1 dB ROSNR penalty. In literature, there exist numer-

Table 4.4: Back-to-back ROSNR for a BER of 10^{-9} at a symbol rate of 10 GBd.

modulation format	ROSNR in dB
NRZ-OOK	13.7
RZ-OOK	12.9
RZ-DBPSK	9.7
RZ-DQPSK	14.6

ous other definitions of the NLT, where it is also defined with respect to eye-opening penalty, Q-factor penalty, different values of ROSNR penalty and OSNR margin. The NLT is a very convenient measure since it is independent of the bit rate (i.e. the required back-to-back OSNR) and it ensures that the nonlinear perturbation δ_{NL} [Eq. (3.10)] is small enough to verify the condition of a weakly nonlinear regime. Furthermore, it gives the maximum reasonable launch power into a system, since increasing the launch power beyond the NLT would not yield a significantly larger OSNR margin at the receiver due to additional nonlinear distortion of the signal.

4.2 Numerical Verification of the Single-Span Model

This section presents a verification of the single-span model by means of numerical simulation. The simulation model described in the previous section is applied to provide more detailed verification than already published in previous work [124, 125, 165–167]. First, the performance of a single-span system in terms of NLT is determined. Subsequently it is tested to what extent these single-span results are applicable to multi-span systems with different dispersion maps. Starting from a resonant dispersion map with full inline dispersion compensation per span, the common scheme of single-periodic maps with constant or randomly varying RDPS is also considered.

4 Application to the Design of Fibre-Optic Transmission Systems

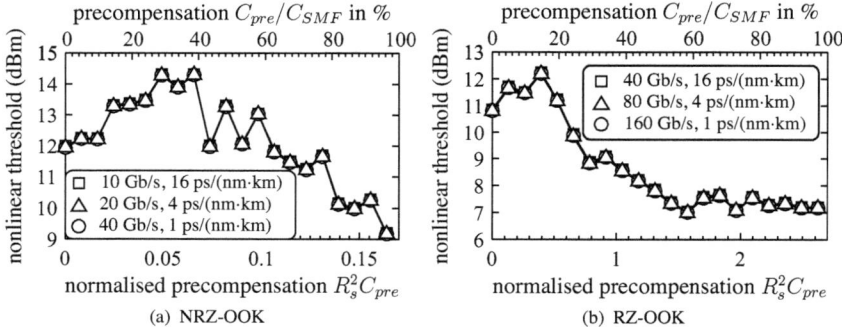

Figure 4.6: Nonlinear threshold as a function of normalised precompensation for transmission of five WDM channels over a single span of fibre with (a) NRZ-OOK and spectral efficiency $S = 0.2$ b/s/Hz at normalised dispersion $C_1 = 0.0445$, (b) RZ-OOK and $S = 0.4$ b/s/Hz at $C_1 = 0.71$.

4.2.1 Single-Span Systems

According to Eq. (3.25), the NLTF for transmission over a single span with dispersion precompensation does depend on only three parameters: its 3-dB bandwidth Ω_s, the amount of precompensation and the ratio of nonlinear coefficient to attenuation coefficient. Evaluation of the NLTF at $\Delta\Omega = R_s^2$ results in

$$\eta(R_s^2) = \frac{\gamma}{\alpha} \frac{1}{1 - j\mathrm{sgn}(\beta_2)C_1} \exp\left(jR_s^2 C_{pre}\right). \tag{4.1}$$

Assuming a constant spectral efficiency, it can be conjectured from this equation that system configurations with equal normalised dispersion C_1 [Eq. (3.45)] and normalised precompensation $R_s^2 C_{pre}$ have equal NLT. This is verified using two exemplary configurations of practical interest: 5×10 Gb/s NRZ-OOK over SSMF with 50 GHz channel spacing ($S = 0.2$ b/s/Hz) and 5×40 Gb/s RZ-OOK over SSMF with 100 GHz channel spacing ($S = 0.4$ b/s/Hz). The first configuration is typical for 10 Gb/s-based legacy systems, while the second constitutes a possible upgrade path to 40 Gb/s. Furthermore, these two system configurations represent transmission in two different regimes. While the first system is mainly impaired by interchannel nonlinearities, the latter operates in the pseudo-linear transmission regime, where intrachannel nonlinearities limit the NLT.

Fig. 4.6 shows the NLT for transmission over 80 km fibre. The normalised dispersion

4.2 Numerical Verification of the Single-Span Model

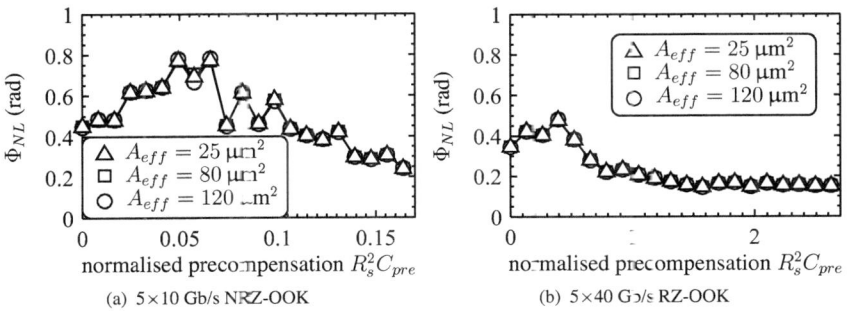

Figure 4.7: Nonlinear phase shift Φ_{NL} for P_0 = NLT (in units Watt) as a function of normalised precompensation. The signal is transmitted over 80 km fibre with dispersion parameter $D = 16$ ps/(nm·km) and varying effective core area. (a) 5×10 Gb/s NRZ-OOK with spectral efficiency $S = 0.2$ b/s/Hz, (b) 5×40 Gb/s RZ-OOK with $S = 0.4$ b/s/Hz.

C_1 is kept constant, while bit rate and GVD are varied. In each case, the precompensation is varied from 0% to 100% of the fibre's cumulated dispersion C_{SMF}. The results indicate that the NLT is indeed determined by normalised dispersion C_1 and normalised precompensation $R_s^2 C_{pre}$ only. Please note that the binary sequences assigned to the WDM channels are delayed by an arbitrary number of bits. This number is kept constant for all simulation points. The NLT therefore represents only one specific alignment of bit pattern in nonlinearly interacting wavelength channels. In order to estimate the true NLT one would have to simulate many configurations with randomly chosen delays and carrier phase constellations. From these runs a time-averaged BER can be determined. However, this would be extremely time consuming. Since the conclusion holds for any random values of delay and carrier phase alignment, it is also valid for the NLT determined by a time-averaged BER obtained by many simulation runs.

In above simulations, only the dispersion parameter of the fibre is varied, while nonlinear coefficient and fibre loss are kept constant. However, dispersion parameter and effective core area of a fibre are not entirely independent parameters. Usually, a reduced dispersion parameter necessitates a smaller effective core area. According to Eq. (2.16), the nonlinear coefficient of a fibre is inversely proportional to the effective core area. Thus, a reduced dispersion parameter has to be traded with a larger nonlinear coefficient and consequently larger nonlinear phase shift at equal input power. Besides normalised dispersion and precompensation, the nonlinear phase shift Φ_{NL}

63

4 Application to the Design of Fibre-Optic Transmission Systems

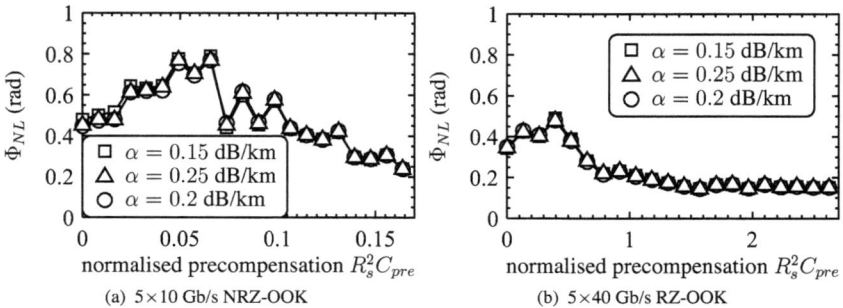

(a) 5×10 Gb/s NRZ-OOK

(b) 5×40 Gb/s RZ-OOK

Figure 4.8: Nonlinear phase shift Φ_{NL} for $P_0 = $ NLT (in units Watt) as a function of normalised precompensation. The signal is transmitted over 80 km fibre with varying fibre loss. The dispersion parameter is adjusted in each case to keep C_1 constant. (a) 5×10 Gb/s NRZ-OOK with spectral efficiency $S = 0.2$ b/s/Hz, (b) 5×40 Gb/s RZ-OOK with $S = 0.4$ b/s/Hz.

is the third defining parameter for a single-span system. Taking above examples of 10 Gb/s and 40 Gb/s systems, it is verified that the allowed nonlinear phase shift for 1 dB ROSNR penalty remains constant when the effective core area is varied. Apart from an effective core area $A_{eff} = 80$ µm^2, which is common for SSMF, very small (25 µm^2) and large (120 µm^2) effective core areas are considered. The results are shown in Fig. 4.7. Indeed, the nonlinear phase shift for $P_0 = $ NLT (in units Watt) is independent of the effective core area.

So far, the fibre loss has been assumed to be $\alpha = 0.2$ dB/km. However, legacy fibres often have larger losses, while modern fibres achieve losses below that value. For constant C_1, the nonlinear phase shift resulting in 1 dB ROSNR penalty should remain constant when the fibre loss is varied, i.e. the NLT is inversely proportional to fibre loss. In order to keep C_1 constant, the dispersion parameter of the fibre is varied. It is $D = 12, 16, 20$ ps/(nm·km) for $\alpha = 0.15, 0.2, 0.25$ dB/km. The nonlinear phase shift for 10 Gb/s NRZ-OOK and 40 Gb/s RZ-OOK transmission over 80 km fibre is shown in Fig. 4.8. The nonlinear phase shift is verified to be invariant to variations of fibre loss for constant C_1.

According to the above results, it is possible to simulate a single-span system with just a single symbol rate and a single value of the nonlinear coefficient with varying dispersion parameter and precompensation to fully describe the obtainable NLT for arbitrary symbol rate and fibre type. Results of such a simulation are shown in Fig. 4.9, where the NLT (contour line labels) is plotted for single-channel and

4.2 Numerical Verification of the Single-Span Model

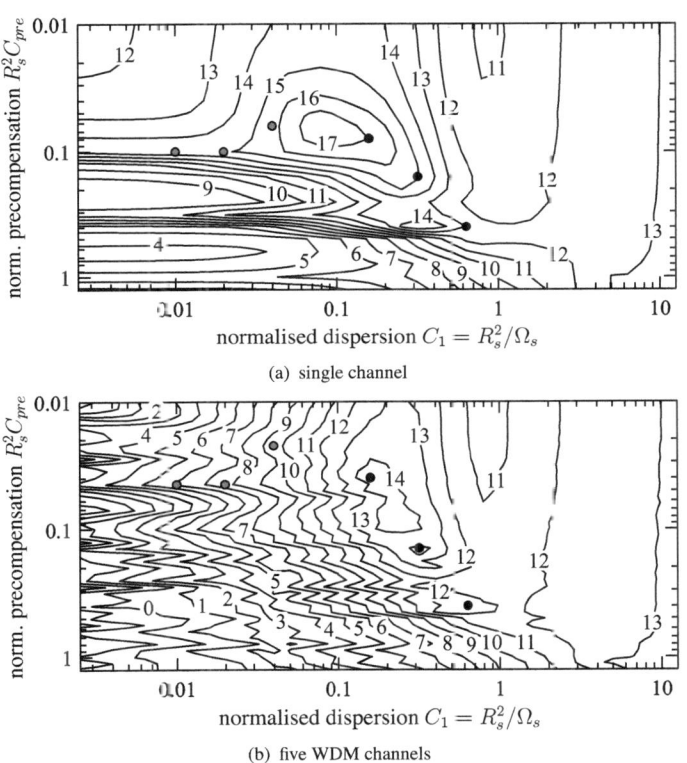

Figure 4.9: Nonlinear threshold in dBm (contour line labels) for RZ-OOK modulation and transmission of (a) a single channel with symbol rate R_s and (b) a $5 \times R_s$ WDM signal with 0.4 b/s/Hz spectral efficiency over a single span with precompensation C_{pre}. Circles indicate configurations giving maximum nonlinear threshold for dispersion parameter $D = 4, 8, 16$ ps/(nm·km) and a bit rate of 10 Gb/s (grey dots) or 40 Gb/s (black dots).

4 Application to the Design of Fibre-Optic Transmission Systems

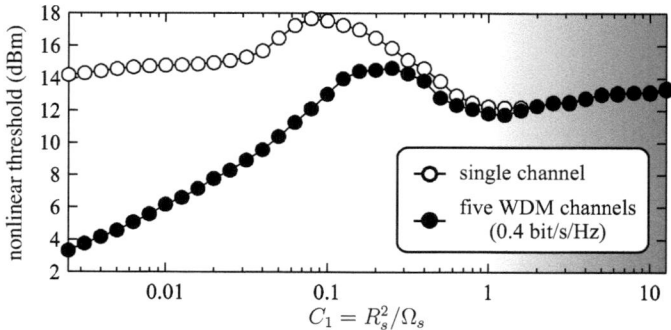

Figure 4.10: Nonlinear threshold for optimised normalised precompensation $R_s^2 C_{pre}$ in case of single-channel (white circles) and $5 \times R_s$ WDM (black circles) transmission. Due to insufficient bit-sequence length, the nonlinear threshold is likely to be overestimated in the shaded area, i.e. for large channel memory.

WDM transmission of RZ-OOK versus normalised dispersion C_1 and precompensation $R_s^2 C_{pre}$. The lower bound of the abscissa corresponds to a bit rate of 10 Gb/s and a fibre dispersion of $D = 1$ ps/(nm·km), while the upper bound corresponds to 160 Gb/s and $D = 18$ ps/(nm·km). This plot completely describes the performance of single-span transmission for arbitrary bit rate and fibre type for 33% RZ-OOK. However, in case of WDM transmission the results are valid for constant spectral efficiency only. The impact of varying spectral efficiency is discussed towards the end of this chapter in section 4.3.2. Furthermore, the NLT for system configurations mainly impaired by interchannel nonlinearities holds only for the specific simulated set of bit pattern timings. Neither does it reflect the true best or worst case, nor does it allow to draw a conclusion about the NLT for a time-averaged BER. The influence of the statistics of interchannel nonlinearities on the NLT is discussed in greater detail in the context of DPSK formats in section 4.3.1.

The NLT obtained for optimised precompensation is shown in Fig. 4.10. It compares the NLT of single-channel transmission with that of WDM transmission at a spectral efficiency of 0.4 b/s/Hz. The NLT for $C_1 > 1$ (shaded region) is likely to be overestimated, since the length of the simulated binary sequence is not sufficient to correctly account for all interacting bit patterns of a single wavelength channel. For example, with Eq. (2.19) and $\Delta\omega_s = 2\pi R_s$ the number of overlapping symbols[2] at $C_1 = 10$

---
[2]Since the bandwidth containing 90% of the signal power is $\Delta\omega_s = 4\pi R_s$ [168], this can be viewed as a low estimate for the number of overlapping symbols.

4.2 Numerical Verification of the Single-Span Model

is $m \approx 64$. Thus, intrachannel nonlinear distortions are underestimated [46]. Recent numerical results with sequence lengths up to 2^{17} indicate that the NLT does not increase for larger values of C_1 [146]. Further note should be taken that the results represent the NLT due to nonlinear impairments in the absence of net residual dispersion. In certain scenarios, the NLT can be further increased by careful optimisation of the net residual dispersion, e.g. in SPM-limited systems (single-channel transmission at small C_1 in Fig. 4.10) [49].

A comparison between single-channel and WDM transmission reveals the severe impact of FWM for small C_1, where the NLT for WDM transmission can be up to 10 dB lower than in the single-channel case. In contrast, there is negligible penalty due to interchannel nonlinear effects for sufficiently large bit rate and/or dispersion ($C_1 \geq 0.3$), i.e. in the pseudo-linear regime [29]. The maximum NLT is achieved in a trade-off between inter- and intrachannel nonlinear effects at $C_1 = 0.25$.

As discussed in section 4.1.4, the OSNR margin at the receiver is an important criterion for the design of a fibre-optic transmission system. Knowledge of the NLT for optimised precompensation as shown in Fig. 4.10 enables simple estimation of the achievable OSNR margin. In the following, strategies to maximise the OSNR margin are discussed on the basis of a simple single-span example. Although the OSNR margin is usually not of concern in a single-span system, this discussion allows to illustrate the main principles. A common design decision faced by engineers designing a fibre-optic transmission system concerns the quest for the best combination of fibre type and symbol rate. In order to answer that question for single-span transmission of RZ-OOK, three fibre types are considered: SLAF, SSMF and NZDSF. Dispersion parameter D_{SMF} and fibre loss of these fibres are summarised in Tab. 4.3. The length of the transmission fibre is $L_{SMF} = 100$ km and the EDFA noise figure is $F_{OA} = 6$ dB. The OSNR at the receiver of a transmission link with single-stage amplification is then determined by Eq. (2.3), with $\alpha_{DCF} = 0.5$ dB/km and $L_{DCF} = -L_{SMF}D_{SMF}/D_{DCF}$ for complete inline dispersion compensation [3]. The launch power P_{in} equals the NLT shown in Fig. 4.10. To avoid ambiguities due to insufficient DBBS length, only NLTs for $C_1 \leq 1$ are considered. The ROSNR is 1 dB higher than the value given in Tab. 4.4 due to operating at the NLT.

The resulting OSNR margin is shown in Fig. 4.11(a). The maximum OSNR margin is achieved by using SLAF and a symbol rate of about 16 GBd, because this

---

[3] The dispersion parameter of the DCF is assumed to be $D_{DCF} = -100$ ps/(nm·km).

4 Application to the Design of Fibre-Optic Transmission Systems

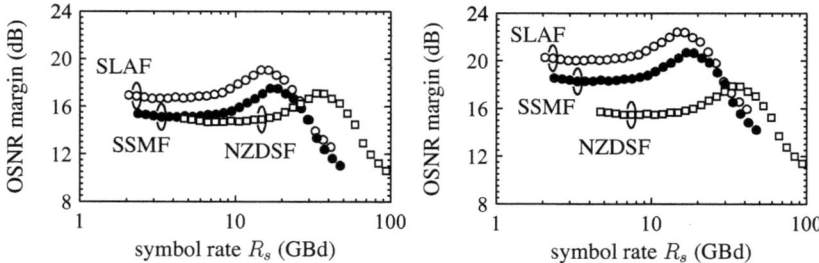

Figure 4.11: OSNR margin as a function of symbol rate for transmission of $5 \times R_s$ RZ-OOK with a spectral efficiency of 0.4 b/s/Hz over 100 km fibre with optimised precompensation and (a) single-stage amplification or (b) dual-stage amplification.

configuration maximises the NLT while minimising the required received OSNR under the given constraints. When transmitting at bit rates per WDM channel higher than 30 Gb/s, NZDSF becomes the best choice. This is due to an increasing impact of intrachannel nonlinearities when transmitting over fibres with large dispersion parameter. Furthermore, a large dispersion parameter necessitates a longer DCF to compensate the dispersion. For example, the loss of a DCF compensating for a SLAF is 6.2 dB larger than the loss in case of NZDSF. As discussed in chapter 2, dual-stage amplification is used to mitigate the impact of DCFs on the received OSNR. Assuming $P_0/P_{DCF} = 10$ to ensure linear transmission in the DCF, the received OSNR for dual-stage amplification is determined by Eq.(2.4) and (2.5). The resulting OSNR margin is shown in Fig. 4.11(b). While the OSNR margin increases only slightly for NZDSF (less than 1 dB) it increases by about 3.2 dB for SSMF and 3.3 dB for SLAF. The conclusion remains valid that SLAF enables the overall maximum OSNR margin, while NZDSF is better suited for transmission at high symbol rates. In the next section it will be shown how these results can be extended to systems with multiple spans.

4.2.2 Multi-Span Systems with Single-Periodic Dispersion Map

In the following, the general class of multi-span systems is treated with respect to a more confined, albeit common, scenario. In a system with a single-periodic disper-

4.2 Numerical Verification of the Single-Span Model

sion map, dispersion is compensated per span by a certain amount. If dispersion is fully compensated per span (i.e. no RDPS), this compensation scheme is referred to as full inline dispersion compensation. First, multi-span systems with such a so-called resonant dispersion map are discussed. The next section goes one step further by introducing a constant amount of RDPS. Last but not least, the impact of small random variations in RDPS, which are inevitable in practical systems, is quantified. For each system configuration, it is explored to what extent multi-span systems can be approximated by single-span systems.

Resonant Dispersion Map

A resonant dispersion map is characterised by the absence of RDPS. Cumulated dispersion at the input of each span equals the amount of precompensation introduced at the transmitter. Hence, signal waveforms are the same at the input of each span, except for nonlinear signal distortions. According to Eq. (3.10), the nonlinear perturbation generated in each span will thus be very similar and the magnitude of the nonlinear perturbation should accumulate approximately in a linear fashion with the number of spans. If this is the case, input power per channel has to be scaled such that the cumulated nonlinear phase shift [Eq. (3.35)] remains constant. In order to test this hypothesis, WDM transmission over five, ten and twenty spans with full inline dispersion compensation per span is compared to transmission over a single span.

The results are shown in Fig. 4.12, where the nonlinear phase shift Φ_{NL} for $P_0 =$ NLT (in units Watt) is plotted versus normalised precompensation. For the most part, these results indicate that the NLT found in single-span transmission can indeed be generalised for systems with an arbitrary number of spans. However, the nonlinear phase shift for transmission of 5×10 Gb/s NRZ-OOK, which is predominantly impaired by interchannel nonlinearities, shows deviations of up to 0.2 rad between single-span and multi-span results [Fig. 4.12(a)]. Just recently, a group from the Università degli Studi di Parma reported similar results obtained by averaging models [169]. They determined the nonlinear phase shift resulting in 3 dB ROSNR penalty for NRZ-OOK with span counts up to 100. While dispersion precompensation was optimised, full inline dispersion compensation was used. It was reported that the allowable nonlinear phase shift is independent of the number of spans for bit rates and GVD above 20 Gb/s and 8 ps/(nm·km) (i.e. $C_1 > 0.09$).

4 Application to the Design of Fibre-Optic Transmission Systems

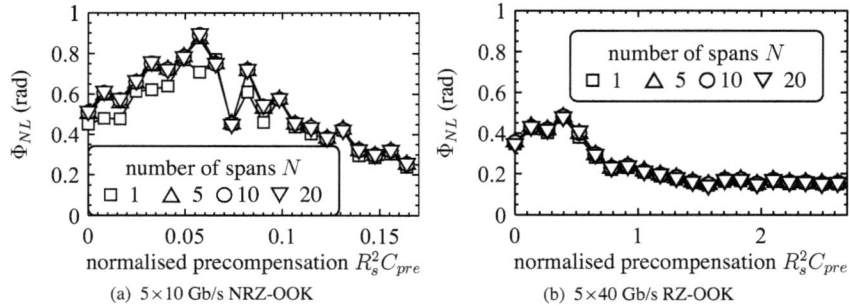

Figure 4.12: Nonlinear phase shift Φ_{NL} for $P_0 = $ NLT (in units Watt) as a function of normalised precompensation for transmission over N spans of SSMF. (a) 10 Gb/s NRZ-OOK with spectral efficiency $S = 0.2$ b/s/Hz, (b) 40 Gb/s RZ-OOK with $S = 0.4$ b/s/Hz.

The smaller C_1 becomes, the more the allowable nonlinear phase shift diverges for different span count. However, the allowable nonlinear phase shift of the single-span system always constitutes the lower bound.

Fig. 4.12(b) presents the results for an exemplary system mostly limited by intra-channel nonlinearities. The perfect match of results for transmission over different numbers of spans verifies that the NLT is reduced with increasing number of spans, thus keeping the cumulated nonlinear phase shift constant. This scaling behaviour was also verified for 160 Gb/s single-channel transmission of Gaussian pulses by means of numerical simulations [127].

These results confirm that the equivalent single-span model derived in chapter 3 is applicable to arbitrary multi-span systems with full inline dispersion compensation per span. Thus, Fig. 4.10 does not only completely describe the maximum NLT of RZ-OOK in single-span systems but it also covers this additional class of multi-span systems.

Constant Residual Dispersion per Span

Introduction of RDPS opens up a new degree of freedom in dispersion-map design. According to Eq. (3.31) and (3.29) the NLTF is invariant to an introduction of RDPS, as long as dispersion precompensation is adjusted properly. However, the approximation used in the derivation of Eq. (3.31) is only valid under the condition that

4.2 Numerical Verification of the Single-Span Model

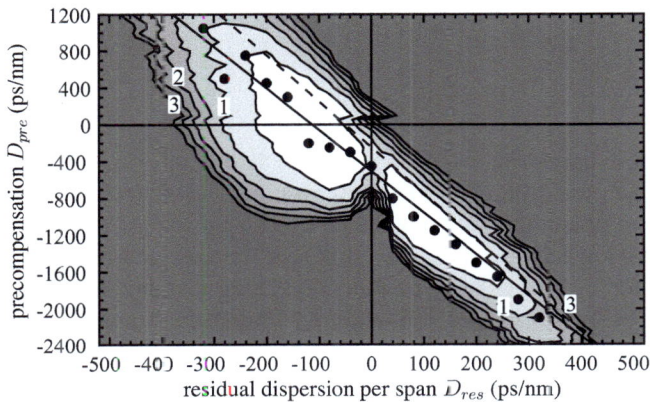

Figure 4.13: ROSNR penalty in dB (contour line labels) as a function of dispersion map parameters for transmission of 5×10 Gb/s NRZ-OOK over 10×80 km SSMF with 4 dBm launch power and spectral efficiency 0.2 b/s/Hz. Black dots indicate numerically obtained optimum precompensation for each value of RDPS. The solid line represents optimum precompensation as a function of RDPS obtained semi-analytically with the equivalent single-span model. Optimum precompensation of the single-span system C'_{pre} is determined numerically [cp. Fig.4.6(a)]. Then Eq. (3.29) is used to calculate optimum precompensation for the multi-span system. Dashed line represents analytically predicted optimum precompensation as a function of RDPS according to Killey's formula [Eq. (3.55)].

RDPS is small [Eq. (3.30)]. The RDPS in practical systems may well exceed the required small values and thus violate this condition. Can knowledge about the NLT of a single-span system still be of value for the design and optimisation of single-periodic dispersion maps? This question is the main focus of this section.

The equivalent single-span model represents an approximation of the cumulated non-linear perturbation at the receiver. Although approximation of the NLTF's magnitude becomes inaccurate for large RDPS, single-span optimisation can be used to obtain optimum dispersion map parameters. Having found the optimum value of dispersion precompensation for the single-span system, optimum precompensation as a function of RDPS and span count is determined by Eq. (3.29). For example, points of maximum NLT for single-span transmission with a bit rate of 40 Gb/s and a fibre dispersion parameter of $D = 4, 8, 16$ ps/(nm·km) are indicated by black dots in Fig. 4.9(a). With Eq. (3.29), the rule for optimum precompensation [Eq. (3.56)] of

4 Application to the Design of Fibre-Optic Transmission Systems

multi-span systems is reproduced and the value of K is found to be

$$K \approx -\frac{D}{\alpha}\ln\left(\frac{2}{1+\exp(-\alpha L)}\right). \tag{4.2}$$

This is in good agreement with the results of Killey in [59]. To test the accuracy of this approach for dispersion map optimisation, it is compared to full numerical optimisation of precompensation and RDPS. The considered transmission line consists of 10×80 km SSMF spans. Two different transmission regimes are examined. Systems limited by interchannel nonlinearities are accounted for by transmission of 5×10 Gb/s NRZ-OOK, while limitation by intrachannel nonlinearities occurs for transmission of 5×40 Gb/s RZ-OOK. As in the previous section, channel spacing is chosen as 50 GHz and 100 GHz to conform with ITU recommendations. This results in a spectral efficiency of 0.2 b/s/Hz and 0.4 b/s/Hz for NRZ-OOK and RZ-OOK, respectively. Launch powers correspond to the NLT for single-span transmission with optimised precompensation according to Fig. 4.6 (i.e. 14.3 dBm for NRZ-OOK and 12.2 dBm for RZ-OOK).

The ROSNR penalty as a function of dispersion precompensation and RDPS for 10 Gb/s NRZ-OOK transmission with 4 dBm launch power is shown as a contour plot in Fig. 4.13. The contour line labels denote ROSNR penalty in dB. The solid strait line represents optimum precompensation as a function of RDPS according to Eq. (3.29). Compared to the numerical results, the deviation in predicted optimum precompensation within the contour for 1 dB ROSNR penalty is less than 250 ps/nm and the difference in ROSNR penalty is less than 0.19 dB. This adequate accuracy makes the equivalent single-span model suited for dispersion-map optimisation in systems mainly impaired by interchannel nonlinearities. The dashed line indicates predicted optimum precompensation as a function of RDPS according to Eq. 3.55, which was derived in the context of 40 Gb/s transmission. Deviations in predicted optimum precompensation up to 700 ps/nm and differences in ROSNR penalty up to 0.6 dB show that it is indeed not designed for this transmission regime.

Fig. 4.14 plots the ROSNR penalty for RZ-OOK transmission with a launch power of 2 dBm per channel. For the optimum dispersion map, a minimum ROSNR penalty of 0.7 dB is achieved. For configurations inside the contour of 1 dB ROSNR penalty, the maximum mismatch between analytically and numerically obtained optimum precompensation is 40 ps/nm in case of the equivalent single-span model and 72 ps/nm when using Eq. (3.55). Analytically found optima result in additional ROSNR penal-

4.2 Numerical Verification of the Single-Span Model

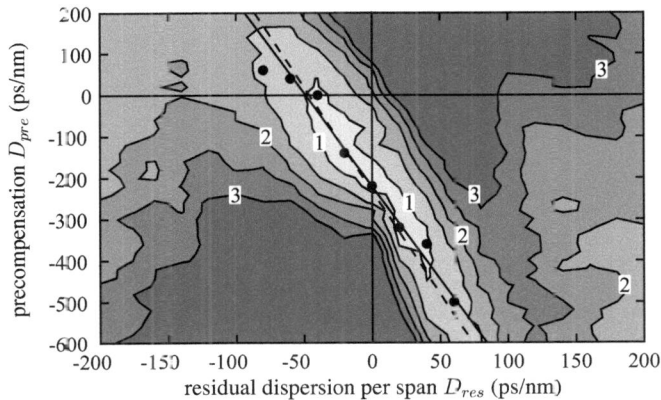

Figure 4.14: ROSNR penalty in dB (contour line labels) as a function of dispersion map parameters for transmission of 5×40 Gb/s RZ-OOK over 10×80 km SSMF with 2 dBm launch power and spectral efficiency 0.4 b/s/Hz. Black dots indicate numerically obtained optimum precompensation for each value of RDPS. Solid and dashed line represent predicted optimum precompensation as a function of RDPS according to the equivalent single-span model and Killey's formula [Eq. (3.55)], respectively.

ties of 0.08 dB and 0.32 dB for equivalent single-span model and Killey's model, respectively. These results indicate that the equivalent single-span model is suited for dispersion map optimisation in WDM OOK systems and offers comparable accuracy to already established models.

For both simulated system configurations, i.e. in systems limited by interchannel as well as intrachannel nonlinearities, the ability of the equivalent single-span model to make predictions about optimum dispersion maps is verified. But what about the equivalence of single-span and multi-span system in terms of ROSNR penalty? According to the approximated NLTF in Eq. (3.31) all system configurations having equal equivalent precompensation C'_{pre} should generate the same nonlinear perturbation and thus lead to the same ROSNR penalty. One can see at first glance in Fig. 4.13 and 4.14 that this is not the case for ROSNR penalty along the solid line and large values of RDPS. Fig. 4.15 shows the ROSNR penalty along the solid line for both NRZ-OOK and RZ-OOK. The light grey area represents the validity region according to Eq. (3.36). While the configuration limited by intrachannel nonlinearities does indeed show similar ROSNR penalty inside the validity region, this is not the case for the 10 Gb/s system limited by interchannel nonlinearities. In contrast, results for single-channel 10 Gb/s NRZ-OOK transmission over 5 and 15 spans re-

4 Application to the Design of Fibre-Optic Transmission Systems

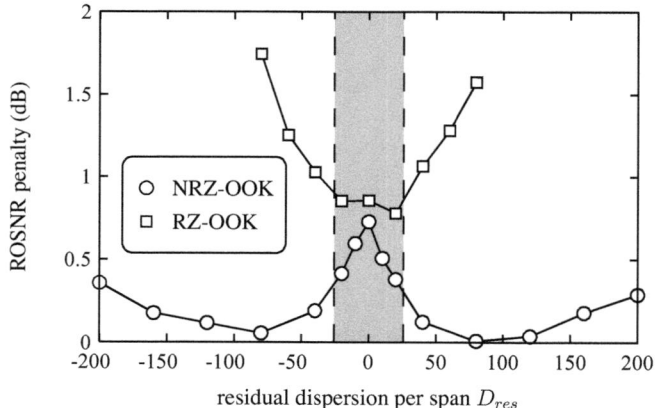

Figure 4.15: ROSNR penalty as a function of RDPS for transmission of 5×10 Gb/s NRZ-OOK and 5×40 Gb/s RZ-OOK over 10 spans of SSMF. The equivalent precompensation is kept constant. This corresponds to the ROSNR penalty along the solid lines in Fig. 4.13 and 4.14. The grey area represents the validity region of the equivalent single-span model according to Eq. (3.36).

ported in [121], indicate a validity of the equivalent single-span model for RDPS at least up to 150 ps/nm (5 spans) and 50 ps/nm (15 spans). Apparently, the equivalent single-span model is able to correctly describe SPM-limited transmission but can not accurately predict the influence of RDPS on the impact of interchannel nonlinearities.

Closely linked to the estimation of ROSNR penalties is the estimation of expected NLT for multi-span transmission. According to the equivalent single-span model, nonlinear coefficient, attenuation coefficient and NLT for single-span transmission define a maximum allowable nonlinear phase shift. The NLT for multi-span transmission is then determined by Eq. (3.35) and is inversely proportional to the number of transmission spans N. As shown in the previous section this holds with good accuracy for multi-span systems with a resonant dispersion map. However, RDPS is known to reduce the resonant accumulation of nonlinear distortions along a transmission line. It is therefore expected that the NLT is increased compared to the resonant case. To quantify this increase, the NLT is determined for transmission of 5×10 Gb/s NRZ-OOK and 5×40 Gb/s RZ-OOK over 5, 10 and 20 spans of SSMF and compared to single-span transmission. In each case, precompensation and RDPS are optimised along the straight line defined by the optimum equivalent single-span

4.2 Numerical Verification of the Single-Span Model

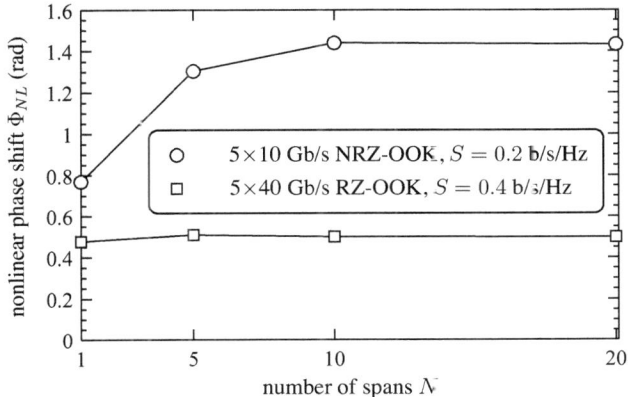

Figure 4.16: Nonlinear phase shift Φ_{NL} for $P_0 = \mathrm{NLT}$ (in units Watt) as a function of span count for transmission over N spans of SSMF with optimised dispersion map.

precompensation and Eq. (3.29). From Fig. 4.6, the optimum values for equivalent precompensation are $D'_{pre} = -512$ ps/nm for 10 Gb/s and $D'_{pre} = -192$ ps/nm for 40 Gb/s. Please note that these are not the exact optimum values, since the step width for precompensation in Fig. 4.6 is quite large (64 ps/nm). The resulting allowed nonlinear phase shift for each number of spans is shown in Fig. 4.16. The NLT for NRZ-OOK transmission is mainly limited by XPM. Optimising the RDPS leads to a reduced accumulation of XPM-induced distortions and thus allows for a larger cumulated nonlinear phase shift. For ten or more spans the allowed nonlinear phase shift almost doubles compared to the single-span case. When interchannel nonlinearities are the dominant source of nonlinear impairments, the NLT of the single-span system can therefore only act as a lower bound for the expected NLT of a multi-span system with optimised dispersion map. In contrast to that, the NLT for RZ-OOK transmission at 40 Gb/s is limited by intrachannel nonlinearities. Although an optimisation of the RDPS allows for an increased cumulated nonlinear phase shift, this increase is less pronounced than for 10 Gb/s transmission. In this case, the single-span NLT indeed constitutes a good estimate for the NLT of multi-span systems with optimised dispersion map.

The equivalent single-span model presents a simple way to estimate expected performance not only in single-span systems but also in dispersion-managed multi-span systems with a single-periodic dispersion map. Furthermore, the dispersion map of

4 Application to the Design of Fibre-Optic Transmission Systems

a given multi-span system can be easily optimised by using knowledge about the single-span system. However, due to several reasons, the RDPS can not be controlled exactly in practice. Therefore, the ideal dispersion map does not exist in real systems. The next section verifies the equivalent single-span model for the more practical case of systems with small random variations of the RDPS and analyses their effect on ROSNR penalty.

Randomly Varying Residual Dispersion per Span

Due to length variations of the transmission fibre in each span and granularity of commercially available DCMs, the nominal RDPS can usually not be guaranteed throughout a transmission line. It is therefore important to estimate the consequences of an imperfect dispersion map on transmission performance [170, 171]. In this example, transmission of 5×40 Gb/s RZ-OOK over a transmission line consisting of 10 spans is considered. The precompensation D_{pre} is varied from -700 ps/nm to 300 ps/nm in steps of 100 ps/nm. Two different uniform distributions of the RDPS are simulated:

1. mean RDPS: $|\langle C_{res} \rangle| = 51$ ps^2, width $C_{DCM} = 102$ ps^2 (corresponding to a DCM-granularity of 5 km SSMF),

2. mean RDPS: $|\langle C_{res} \rangle| = 102$ ps^2, width $C_{DCM} = 204$ ps^2 (corresponding to a DCM-granularity of 10 km SSMF).

The maximum absolute residual dispersion of an individual span is 80 ps/nm for the first case and 160 ps/nm for the second case. For negative precompensation values, a positive mean RDPS has been assumed and vice versa. This way, all dispersion maps show either a monotonously increasing or decreasing cumulated dispersion at the beginning of each span. A special case occurs for zero precompensation, where the mean RDPS is assumed to be zero. The equivalent precompensation D'_{pre} is determined by Eq. (3.37) and (3.39). For each of the two distributions, 1100 random systems have been simulated. Fig. 4.17 plots the ROSNR penalty versus precompensation of the equivalent single-span system. Penalties for the equivalent single-span system (where $D'_{pre} = D_{pre}$) are shown as a reference (circles). As long as the mean RDPS and the variation of the RDPS are small enough, the qualitative match between single-span systems and random multi-span systems is satisfactory [Fig. 4.17(a)]. For equivalent precompensation in the range -300 ps/nm$\leq D'_{pre} \leq 150$ ps/nm, the

4.2 Numerical Verification of the Single-Span Model

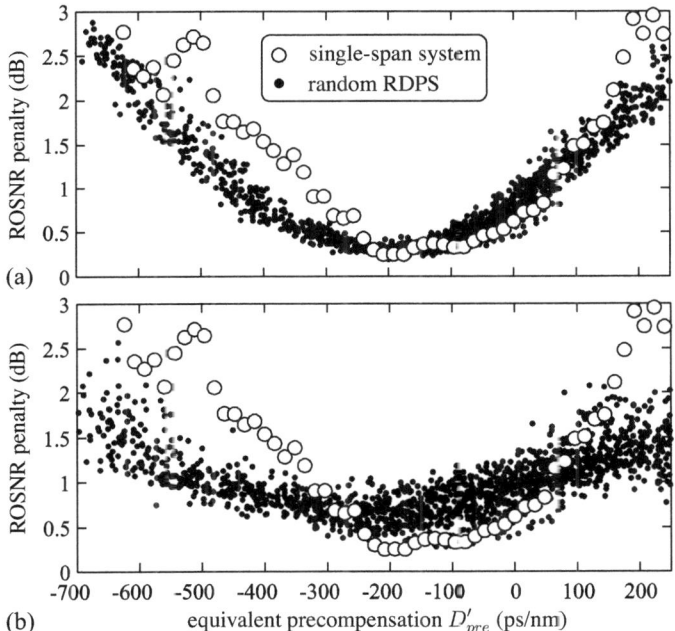

Figure 4.17: ROSNR penalty as a function of equivalent precompensation. Shown are the results of split-step Fourier simulations for transmission of 5×40 Gb/s RZ-OOK over different transmission lines: a single 80 km span of SSMF with launch power of 12 dBm per channel (circles), and 10×80 km SSMF with uniformly distributed residual dispersion per span, varying precompensation and 2 dBm launch power per channel (dots). The distribution parameters are: (a) mean RDPS $\langle C_{res} \rangle = \pm 51$ ps^2, width $C_{DCM} = 102$ ps^2 and (b) mean RDPS $\langle C_{res} \rangle = \pm 102$ ps^2, width $C_{DCM} = 204$ ps^2.

maximum error in ROSNR penalty is less than ± 0.5 dB. Furthermore, the optimum system configuration around a precompensation of about -200 ps/nm is correctly estimated by the single-span system. However, for larger mean RDPS and larger variation per span [Fig. 4.17(b)] approximation by the single-span model becomes too inaccurate and the error in terms of ROSNR penalty can approach 1 dB even near the predicted optimum. The results in Fig. 4.17(b) also indicate that the OSNR margin can be significantly reduced due to DCM granularity, even when special care is taken to match the optimum dispersion map as accurate as possible. It can be concluded that prediction of optimum system parameters by use of the single-span model even holds in the presence of small random variations of RDPS, which inevitably occur in practical systems.

4.3 Implications on the Design of Fibre-Optic Transmission Systems

This section explores some possible applications of the model to the design of fibre-optic transmission systems. With the range of applicability of the equivalent single-span model and its limits established in the previous section, it is now used to provide help with some exemplary design decisions. In section 4.3.1, it is applied in order to compare the performance of binary and quadrature DPSK transmission in the presence of fibre nonlinearity. The impact of an increasing number of encoded bits per transmitted symbol on NLT and OSNR margin is determined. Section 4.3.2 deals with the impact of varying spectral efficiency on the NLT and verifies the duality between the parameters C_1 and C_2 as predicted in section 3.3. Furthermore, it is shown that the NLT in systems with electronic precompensation of intrachannel nonlinearities depends solely on the parameter C_2.

4.3.1 Multilevel Phase-Modulated Signals

So far, only binary OOK modulation has been considered. However, the demand for higher bit rates per WDM channel as well as for an increased spectral efficiency necessitates utilisation of advanced modulation formats such as binary and quadrature DPSK. Compared to binary OOK formats, the main advantage of DBPSK with

4.3 Implications on the Design of Fibre-Optic Transmission Systems

balanced detection is a reduction of the back-to-back ROSNR by about 3 dB [172]. Furthermore, it shows a better tolerance to narrow-band optical filtering [156]. There has been extensive research concerning the tolerance of DPSK formats to fibre nonlinearity. For equal average power, RZ-DPSK formats have half the peak power compared to RZ-OOK formats, because a pulse is sent for every symbol. In most cases, this is the underlying reason for the observed better tolerance towards fibre nonlinearity [159, 173]. In particular, it was shown that in some configurations the detrimental effect of XPM is reduced by employing DBPSK [174–176]. Furthermore, the impact of IFWM, which leads to phase fluctuations and amplitude jitter, is less severe for DPSK formats [177]. This was confirmed by measurements of the NLT for transmission of 40 Gb/s signals over SSMF [178] and other fibre types [179,180]. Numerical simulations showed that DPSK does indeed outperform OOK for a large range of fibre types by even a larger margin than one would suspect from its 3 dB reduction in peak power [181]. This was later supported by experimental comparison of RZ-OOK and RZ-DBPSK transmission over SSMF and dispersion-managed fibre [182].

For the case of M-ary multilevel signals the spectral efficiency defined in Eq. (1.1) is generalised to

$$S = \frac{\log_2(M) R_s}{\Delta f_{ch}}. \tag{4.3}$$

DQPSK transmits two bits per symbol ($M = 4$) and thus enables a doubling of spectral efficiency compared to DBPSK ($M = 2$) for a constant symbol rate and channel spacing [141, 144]. Transmission experiments implementing DQPSK have demonstrated high spectral efficiency without resorting to polarisation-division multiplexing, e.g. [183–185]. Another important advantage of DQPSK is the ability to generate high bit rates while using low-speed electronics. For example, a 100 Gb/s DQPSK signal was generated with a transmitter consisting of electronics specified for 40 Gb/s systems [186]. There are few studies comparing the effect of fibre nonlinearity in DBPSK and DQPSK transmission. One comparison of these two formats for a bit rate of 42.7 Gb/s and equal spectral efficiency of 0.8 b/s/Hz concluded that the nonlinear tolerance is similar in the presence of narrow-band optical filtering [187]. In the following, NLTs of RZ-DBPSK and RZ-DQPSK are compared for single-span WDM transmission. Furthermore, applicability of the equivalent single-span model for dispersion-map optimisation in RZ-DQPSK systems is verified.

In order to compare the tolerance of binary and quadrature RZ-DPSK to fibre non-

4 Application to the Design of Fibre-Optic Transmission Systems

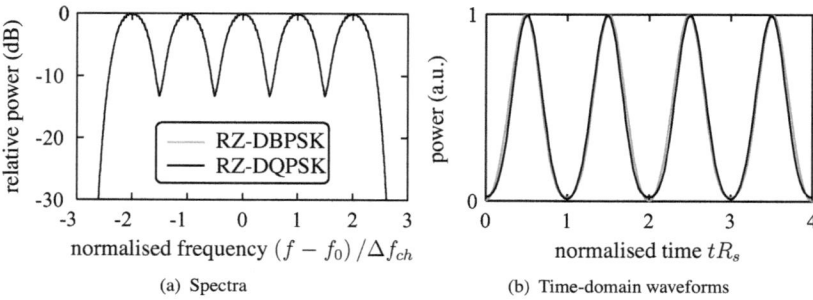

(a) Spectra

(b) Time-domain waveforms

Figure 4.18: (a) Spectra of RZ-DBPSK and RZ-DQPSK for equal channel spacing and symbol rate and (b) associated time-domain waveforms of four exemplary symbols.

linearity, the same single-span investigation as for RZ-OOK in section 4.2.1 is conducted. In the scope of analysis of multilevel modulation formats, the dependence of normalised dispersion C_1 on symbol rate rather than on bit rate becomes important. By choosing equal symbol rate and channel spacing for both binary and quadrature RZ-DPSK, it is ensured that spectra and time-domain waveforms (in the absence of cumulated dispersion) are approximately identical (Fig. 4.18). This implies that quaternary formats have a doubled spectral efficiency compared to binary formats for constant normalised dispersion C_1. For this specific investigation, spectral efficiencies are fixed at 0.4 b/s/Hz and 0.8 b/s/Hz for RZ-DBPSK and RZ-DQPSK, respectively. Since the physical origin of the nonlinear perturbation – the optical power – is distributed in the same way in frequency domain and time domain for both formats, the generated nonlinear perturbation is expected to be almost identical. Nevertheless, the NLT of RZ-DQPSK is reduced compared to RZ-DBPSK due to a reduced distance between signal states in the complex plane.

Fig. 4.19 shows the NLT for single-span transmission of five WDM channels modulated with RZ-DBPSK and RZ-DQPSK, respectively. The curves have the same characteristic shape as for OOK transmission. For $C_1 < 0.3$, where the NLT is limited by interchannel nonlinearities, it increases monotonously with C_1. The penalty induced by FWM strongly depends on the phase relation between the carriers. In a real system, the relative phase between the carriers fluctuates. It is therefore not sufficient to consider only a constant relative phase. Furthermore, the alignment of the transmitted bit sequences in the time domain also affects the penalty induced by FWM and XPM. In order to assess the extent of NLT fluctuation, due to variations

4.3 Implications on the Design of Fibre-Optic Transmission Systems

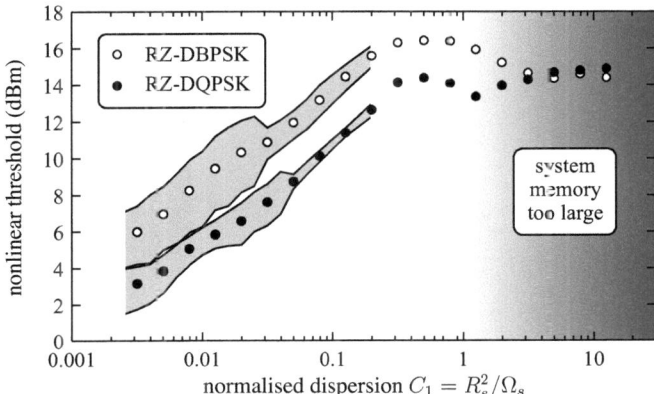

Figure 4.19: NLT for single-span transmission of $5 \times R_s$ RZ-DBPSK and RZ-DQPSK as a function of normalised dispersion C_1. Grey areas are bounded by the respective 10th and 90th percentiles of the NLT distribution resulting from 100 simulation runs with random timings of the transmitted bit sequences in the five WDM channels and random relative carrier phases.

of carrier phases and bit patterns, a statistic approach is adopted. For each value of C_1, 100 simulation runs with random carrier phases and bit pattern alignments were carried out. The precompensation is optimised for maximum mean NLT. The width of the resulting distribution is indicated in Fig. 4.19 by grey areas bounded by the respective 10th and 90th percentiles. Dots represent the mean NLT. At the smallest considered C_1, the difference between the 10th and the 90th percentile amounts to about 2 dB and 4 dB for RZ-DQPSK and RZ-DBPSK, respectively. The maximum observed difference between best and worst case of the conducted 100 simulation runs is 5.9 dB and 5.1 dB for RZ-DBPSK and RZ-DQPSK, respectively. However, such large variations only occur for transmission of low bit rates over fibres with small dispersion parameter, e.g. 4.5 Gb/s over NZDSF with $D = 4.5$ ps/(nm·km). Allowing for less launch power and thus smaller nonlinear perturbation, the variation of NLT for RZ-DQPSK is slightly smaller than for RZ-DBPSK. With increasing C_1 and therewith reduced impact of interchannel nonlinearities, the distribution becomes narrower. Both modulation formats reach maximum NLT at $C_1 \approx 0.4$, where both interchannel and intrachannel nonlinearities contribute to the nonlinear penalty. Further increase of C_1 leads to stronger impact of intrachannel nonlinearities, reducing the NLT. Unfortunately, the computational effort becomes prohibitively large for $C_1 > 1$. According to Eq. (2.19), the required bit-sequence length scales exponen-

4 Application to the Design of Fibre-Optic Transmission Systems

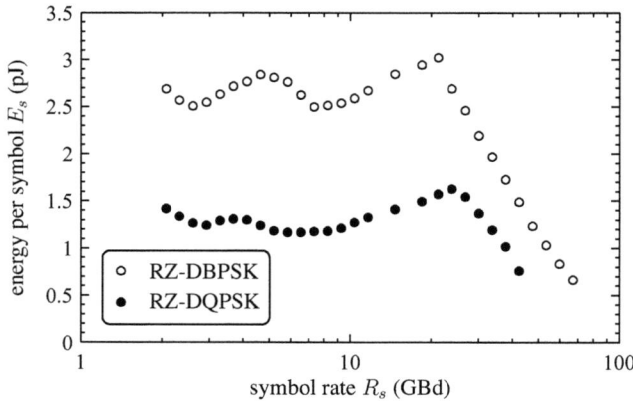

Figure 4.20: Energy per symbol resulting in 1 dB ROSNR penalty as a function of symbol rate for transmission over a single-span of 80 km SLAF with optimised precompensation.

tially with C_1. Due to insufficient length of the employed pseudo-random sequences, the NLT is likely to be overestimated in the shaded region. However, the general trend should be similar to OOK modulation, where recent numerical results indicate that the NLT saturates for large C_1 [146, 169]. Disregarding the uncertain results for $C_1 > 1$, the minimum difference in NLT occurs at optimum $C_1 \approx 0.4$ and amounts to 2 dB. In transmission limited by interchannel nonlinearities, the difference in mean NLT reaches up to 4 dB. A common upgrade scenario is the overlay of RZ-DQPSK channels over existing line infrastructure by just exchanging the terminal equipment. By keeping the channel spacing and the symbol rate constant, the bit rate per channel and the spectral efficiency can be doubled [188]. However, this increase in capacity comes at the cost of a 4.9 dB increase in back-to-back ROSNR (3 dB due to doubled bit rate, 1.9 dB due to a reduced distance between signal states [19]) and at least a 2 dB reduction in NLT. Therefore, the OSNR margin is reduced by about 7 dB in this upgrade scenario, which makes it unfeasible in systems already operating at tight margins.

Besides calculation of the actual OSNR margin (as performed for RZ-OOK in section 4.2.1), there is another way of interpreting the NLT. The energy per symbol resulting in 1 dB ROSNR penalty can be calculated from the results in Fig. 4.19 by

4.3 Implications on the Design of Fibre-Optic Transmission Systems

assuming a specific fibre type. In this case, the maximum energy per symbol is

$$E_s = \frac{\text{NLT}}{R_s}, \qquad (4.4)$$

where the NLT is expressed in units Watt. For M-ary modulation, the energy per bit is obtained as

$$E_b = \frac{E_s}{\log_2 M} \qquad (4.5)$$

The energy per bit is an important measure, since it allows the estimation of channel capacities [9]. Fig. 4.20 shows the energy per symbol as a function of symbol rate for single-span transmission over SLAF. The maximum energy per symbol is obtained at symbol rates of about 21 GBd for RZ-DBPSK and 24 GBd for RZ-DQPSK. Considering the energy per bit, the values for RZ-DQPSK have to be divided by two, according to Eq. (4.5). At the point of maximum energy per bit for RZ-DQPSK, it has about 5.2 dB lower energy per bit than RZ-DBPSK (3 dB due to doubled bit rate of RZ-DQPSK and 2.2 dB reduced NLT). This is in good agreement with the discussion in the previous paragraph.

In order to verify the applicability of dispersion-map optimisation with the equivalent single-span model to multilevel PSK signals, a dispersion map for transmission of five RZ-DQPSK modulated WDM channels over 10 spans is optimised numerically. Three different configurations with 0.8 b/s/Hz spectral efficiency are considered: 10 and 40 GBd over SSMF and 20 GBd over SLAF. According to the single-span results in Fig. 4.19, the last configuration should be least affected by fibre nonlinearity and therefore allow for the highest launch power, while the other two configurations could be employed as upgrades over legacy fibre plant. The launch power is chosen to yield approximately the nonlinear phase shift determined by the NLT of the single-span system. Fig. 4.21 plots the ROSNR penalty as contour lines over the plane spanned by dispersion precompensation and RDPS. The optimum amount of precompensation for each simulated value of RDPS is indicated by black dots. For comparison, optimum precompensation as a function of RDPS as predicted by the equivalent single-span model and by Killey's formula [Eq. (3.55)] are shown as solid and dashed lines, respectively. The 10 GBd system shown in Fig. 4.21(a) and the 40 GBd system shown in Fig. 4.21(b) operate at $C_1 = 0.045$ and $C_1 = 0.71$, respectively. According to Fig. 4.19, the former system is limited by interchannel nonlinearities, while the latter is predominantly affected by intrachannel nonlinearities. The prediction of optimum precompensation yields quite accurate results for both transmission regimes.

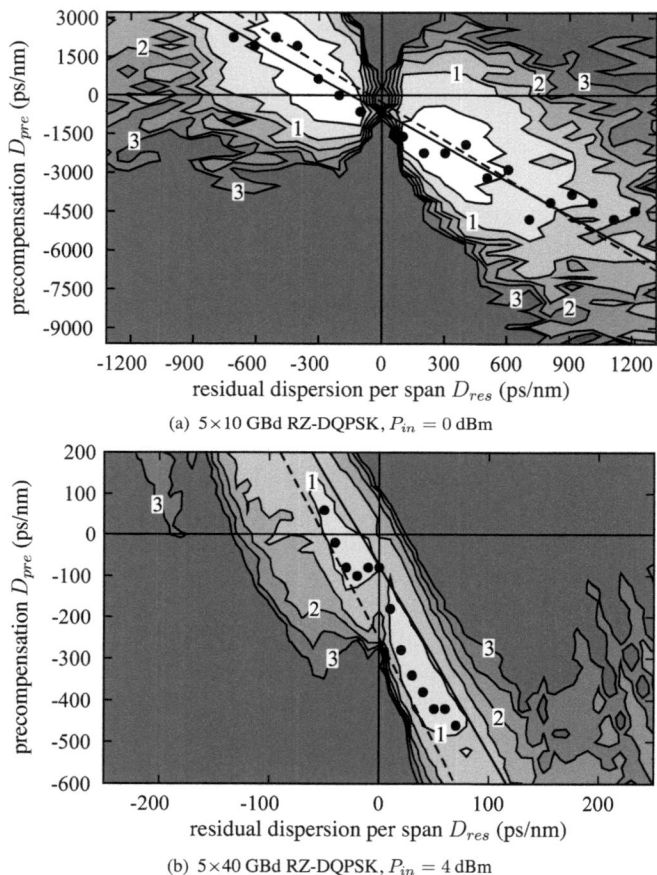

Figure 4.21: Contour plot of ROSNR penalty as a function of dispersion precompensation and RDPS for transmission of (a) 5×10 GBd with $P_{in} = 0$ dBm and (b) 5×40 GBd RZ-DQPSK with $P_{in} = 4$ dBm over 10×80 km of SSMF. Black dots indicate optimum amount of precompensation for each simulated value of RDPS. Solid line is optimum precompensation as a function of RDPS according to Eq. (3.29), where the optimum value of the equivalent precompensation C'_{pre} was numerically obtained. Dashed line is optimum precompensation according to Eq. 3.55.

4.3 Implications on the Design of Fibre-Optic Transmission Systems

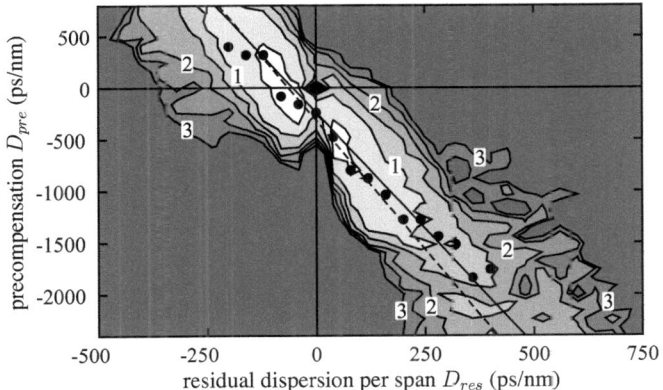

Figure 4.22: ROSNR penalty (contour line labels) as a function of dispersion map parameters for transmission of 5×20 GBd RZ-DQPSK over 10×80 km SLAF with $P_{in} = 4$ dBm.

Fig. 4.22 presents the results for transmission of 20 GBd RZ-DQPSK over SLAF, which constitutes a nearly optimum configuration with respect to NLT and OSNR margin at the receiver. Indeed, a comparison with results of 40 GBd transmission reveals a reduced ROSNR penalty for an optimised dispersion map. This better performance is mainly due to a reduced impact of intrachannel nonlinearities and a smaller nonlinear coefficient of the transmission fibre.

The accuracy of predicting optimum dispersion map parameters with the equivalent single-span model can be estimated by analysing the mismatch between numerically and analytically obtained optimum precompensation as well as the difference in ROSNR penalty that is associated with this mismatch. Fig. 4.23(a) plots the mismatch of normalised precompensation $R_s^2 C_{pre}$ as a function of normalised RDPS $R_s^2 C_{res}$ for symbol rates 10, 20 and 40 GBd. The overall maximum mismatch amounts to 1600 ps/nm, 320 ps/nm and 140 ps/nm for 10, 20 and 40 GBd, respectively. Considering only configurations situated inside the 1 dB penalty contour, the maximum mismatch reduces to 640 ps/nm and 240 ps/nm for 10 and 20 GBd, while it stays the same for 40 GBd. Fig. 4.23(b) plots the corresponding difference in ROSNR penalty. The overall maximum additional ROSNR penalty amounts to 0.76 dB, 0.53 dB and 0.22 dB for 10, 20 and 40 GBd, respectively. Again, considering only configurations inside the 1 dB penalty contour, the additional penalty is reduced to 0.22 dB and 0.06 dB for 10 and 20 GBd while it stays the same for

4 Application to the Design of Fibre-Optic Transmission Systems

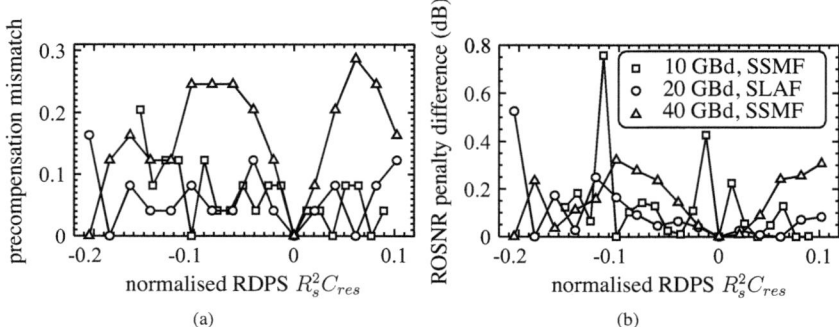

Figure 4.23: (a) Mismatch of normalised precompensation $C_{pre}R_s^2$ and (b) difference in ROSNR penalty between analytically and numerically obtained optima versus normalised RDPS.

40 GBd. Thus, for the considered configurations, a maximum of 0.32 dB additional penalty incurs.

As discussed in section 4.2.2 in the context of 10 Gb/s NRZ-OOK transmission over multiple-spans with optimised dispersion map, the equivalent single-span model does not correctly estimate the NLT and the ROSNR penalty in this transmission regime. Based on the results presented in Fig. 4.21 and 4.22, the ability of the equivalent single-span model to predict ROSNR penalties in multi-span RZ-DQPSK transmission is assessed. For this purpose, Fig. 4.24 plots the ROSNR penalty for optimum equivalent precompensation as a function of RDPS. The region of validity according to Eq. (3.36) is indicated by the light grey area. For the equivalent single-span model to be applicable, the ROSNR penalty should be approximately constant within these boundaries. Similar to NRZ-OOK transmission, the equivalent single-span model does not correctly predict the ROSNR penalty for small C_1, i.e. for 10 GBd ($C_1 = 0.045$) and 20 GBd ($C_1 = 0.23$) in Fig. 4.21. For a symbol rate of 40 GBd ($C_1 = 0.71$), the ROSNR penalty remains remarkably constant within the boundaries and even beyond, thus verifying the equivalent single-span model in this transmission regime. In the same way as for OOK transmission, the accuracy of the equivalent single-span model in predicting actual ROSNR penalties is reduced for small values of C_1. Nevertheless, as discussed previously, the ability to predict optimum precompensation is retained.

The results presented in this section together with results published in [125] indicate that the equivalent single-span model is applicable to DPSK systems, be it binary or quadrature DPSK. However, apart from the considerations above, there is an-

4.3 Implications on the Design of Fibre-Optic Transmission Systems

other important limitation. Due to the nature of its derivation, the equivalent single-span model does not account for nonlinear interactions between signal and noise. Therefore it can not be used to make predictions about systems limited by such interactions, most notably Gordon-Mollenauer phase noise. An example currently of great practical interest is ultra long-haul transmission of phase-modulated signals in conjunction with forward error correction (FEC). In such systems the maintained OSNR during transmission can become very low, leading to strong nonlinear phase noise [30, 119, 176, 189–194]. In order to estimate the impact of nonlinear phase noise for the system configurations considered in this section, a simple method presented in [195] is adopted. The impact of nonlinear phase noise is estimated based on the ratio of the variances of nonlinear phase noise σ_{NL}^2 and linear noise σ_L^2. This ratio is approximately [195]

$$\frac{\sigma_{NL}^2}{\sigma_L^2} = \frac{4}{3}\Phi_{NL}^2, \tag{4.6}$$

where Φ_{NL} is the average nonlinear phase shift according to Eq. (3.35). For $\sigma_{NL}^2/\sigma_L^2 > 1$, nonlinear phase noise is the dominating noise term. For transmission over 10×80 km SSMF with $P_{in} = 0$ dBm [Fig. 4.21(a)], the ratio is $\sigma_{NL}^2/\sigma_L^2 = 0.11$. Increasing the launch power to $P_{in} = 4$ dBm [Fig. 4.21(b)] leads to a ratio of $\sigma_{NL}^2/\sigma_L^2 = 0.69$. For transmission over 10×80 km SLAF with $P_{in} = 4$ dBm (Fig. 4.22) the ratio is $\sigma_{NL}^2/\sigma_L^2 = 0.34$. It can therefore be concluded that the considered system configurations are only weakly affected by nonlinear phase noise. Furthermore, it is conjectured from Eq. (4.6) that the equivalent single-span model should be applicable as long as $\Phi_{NL}^2 \ll 3/4$.

Another implication of the presented results concerns the value of single-span transmission experiments. In conjunction with the equivalent single-span model, measurements of the NLT in simple single-span systems, such as conducted in [178–180], can be used to estimate optimum dispersion map parameters and obtainable NLT in more complex multi-span systems. This is supported by measurements in [178] and [180], where the rule $\text{NLT} + 10 \cdot \log(N) = \text{const}$ was confirmed for 43 Gb/s single-channel transmission of RZ-DBPSK over one and three spans as well as for WDM transmission over one and two spans.

4 Application to the Design of Fibre-Optic Transmission Systems

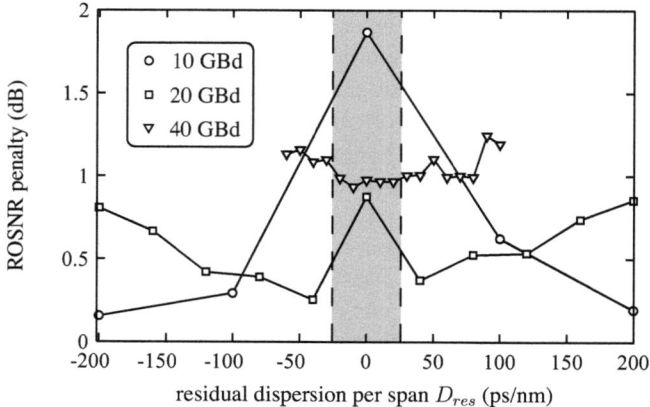

Figure 4.24: ROSNR penalty as a function of RDPS for system configurations shown in Fig. 4.21 and 4.22. The equivalent precompensation is kept constant. This corresponds to the ROSNR penalty along the solid line representing optimum precompensation in those figures. The grey area represents the validity region of the equivalent single-span model according to Eq. (3.36).

4.3.2 Varying Spectral Efficiency

So far, only fixed spectral efficiency has been considered. This section deals with the impact of varying spectral efficiency on the NLT. Considered spectral efficiencies are 0.1, 0.2 and 0.4 b/s/Hz. A further doubling of spectral efficiency to 0.8 b/s/Hz leads to coherent WDM crosstalk between wavelength channels [151]. This crosstalk could distort the results. Therefore, spectral efficiency is limited in this analysis to values not exceeding 0.4 b/s/Hz. The binary sequences transmitted in different channels are decorrelated by delaying them by an arbitrarily chosen number of bits. This number is kept constant for all simulations.

Fig. 4.25(a) plots the NLT for single-span transmission of NRZ-OOK as a function of normalised dispersion C_1. As expected from the considerations presented in section 3.3, the parameter C_1 does not capture the influence of varying spectral efficiency on the NLT. In case of WDM transmission, there are three distinct curves for different spectral efficiency. Only for large symbol rates and dispersion (e. g. for $R_s \geq 40$ GBd and $D \geq 8$ ps/(nm·km)), i.e. for $C_1 \geq 0.3$, the three curves converge and the NLT becomes independent of spectral efficiency. In this regime, nonlinear interaction between wavelength channels is negligible and intrachannel nonlinearities

4.3 Implications on the Design of Fibre-Optic Transmission Systems

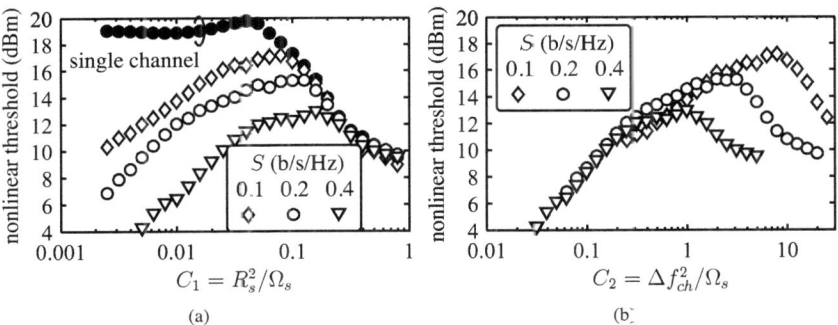

Figure 4.25: Nonlinear threshold for single-span transmission of NRZ-OOK with optimised precompensation and varying spectral efficiency S as a function of (a) normalised dispersion C_1 and (b) normalised dispersion C_2.

are predominantly limiting the NLT (cp. Fig. 3.13). This confirms that the parameter C_1 is only truly universal in systems limited by intrachannel nonlinearities.

When plotting the NLT against normalised dispersion C_2 the curves become invariant to spectral efficiency for system configurations where interchannel nonlinearities limit the NLT, i.e. for $C_2 < 1$ [4] [Fig. 4.25(b)]. However, this normalisation does not capture the impact of intrachannel nonlinearities. Consequently, the curves for systems impaired mostly by intrachannel nonlinearities diverge. Comparison of Fig. 4.25(a) and (b) reveals a dualism between parameters C_1 and C_2. While the normalisation C_1 universally describes scaling of NLT in systems predominantly impaired by intrachannel nonlinearities, the same is true for normalisation C_2 in systems impaired by interchannel nonlinearities. Both normalisations fail when the respective other type of nonlinearity limits the NLT. This supports the conclusions drawn at the end of chapter 3 in the context of FWM and IFWM efficiency. It can be concluded that no single parameter exists, which describes the scaling of NLT for arbitrary bit rate, fiber type and spectral efficiency. However, the possibility to electronically precompensate for intrachannel nonlinearities has emerged in recent years. The remaining nonlinear impairments are caused by interchannel nonlinearities only. It is expected that the normalisation C_2 is universally valid in such systems. This possibility is explored in the next section.

---

[4] Remaining differences in NLT are attributed to non-perfect optimisation of the precompensation. It turned out that the step-size was chosen too large to accurately capture all optima.

4 Application to the Design of Fibre-Optic Transmission Systems

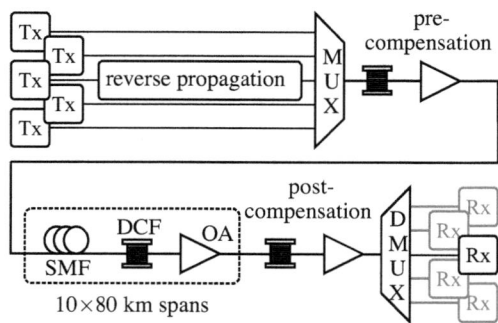

Figure 4.26: Schematic of the system setup with EPD of the centre channel.

Electronic precompensation of intrachannel nonlinearities

Recent advances in electronic digital signal processing brought about the possibility to electronically precompensate for intrachannel nonlinearities in systems with bit rates up to 10 Gb/s per channel [196–201]. Today, the limited sampling rate of digital-to-analogue converters and the amount of channel memory that can be represented by nonlinear filters pose the main limitations for electronic precompensation of intrachannel nonlinearities [201]. With the expected increase in speed of available electronics, it becomes feasible to apply electronic precompensation to systems with per-channel bit rates of 40 Gb/s and higher. Numerical studies of electronic precompensation in 40 Gb/s systems predict an achievable increase in launch power of 8 dB for single-channel [202] and up to 3 dB for WDM configurations [203, 204]. In systems employing electronic predistortion (EPD) of the transmitted signal to precompensate for intrachannel nonlinearities, interchannel nonlinearities are left as the dominant source of nonlinear impairments limiting the channel capacity [21, 203–205]. It is therefore of paramount importance to understand how the influence of interchannel nonlinearities is affected by the choice of bit rate, fibre type and spacing between wavelength channels.

Fig. 4.26 shows the simulation setup for this analysis. In order to remove the impact of intrachannel nonlinearities on the NLT, the centre channel is predistorted by ideal reverse propagation [206, 207]. Limitations due to the sampling rate of digital-to-analogue converters and the amount of channel memory that can be represented by nonlinear filters are neglected. This represents precompensation of intrachan-

4.3 Implications on the Design of Fibre-Optic Transmission Systems

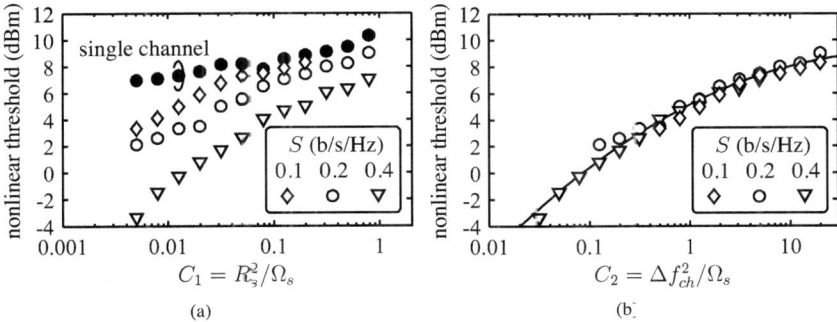

Figure 4.27: Nonlinear threshold for transmission of electronically precompensated NRZ-OOK over 10×80 km spans with optimised precompensation and varying spectral efficiency S as a function of (a) normalised dispersion C_1 and (b) normalised dispersion C_2. The solid line represents a quadratic fit to the data.

nel nonlinearities under ideal conditions. Since OAs are modelled as noiseless for simulation purposes, the remaining nonlinear impairments originate in nonlinear interaction between the wavelength channels. Since the results shown in Fig. 4.25 revealed a lack of accuracy in precompensation optimisation, Kashner's algorithm for a 1-dimensional peak search [208] is implemented with 50 iterations to yield more accurate results for optimum precompensation[5]. Full inline dispersion compensation per span is used.

Fig. 4.27(a) plots the NLT for transmission over 10×80 km spans as a function of parameter C_1. For single-channel transmission with EPD, the NLT is limited by the available multiplexer bandwidth. In this case, the multiplexer filter bandwidth is $2\times R_b$. Although interchannel nonlinearities dominate for WDM transmission, the NLT is invariant to changes of bit rate and GVD for constant C_1 and spectral efficiency. In that sense, C_1 is still universally valid for WDM transmission with EPD, albeit for fixed spectral efficiency only. However, as in systems without EPD, this parameter can not capture the influence of varying spectral efficiency. This is solved by introducing the normalisation C_2. The NLT as a function of C_2 is shown in Fig. 4.27(b). Plotted versus the parameter C_2, the NLT becomes invariant to changes of spectral efficiency. Thus, the parameter C_2 indeed characterises the impact of interchannel nonlinearities for arbitrary bit rate, channel spacing, spectral efficiency

---

[5]This was inspired by discussions with Oscar Gaete who presented an iterative numerical approach to multidimensional optimisation at the ECOC 2008 [209].

4 Application to the Design of Fibre-Optic Transmission Systems

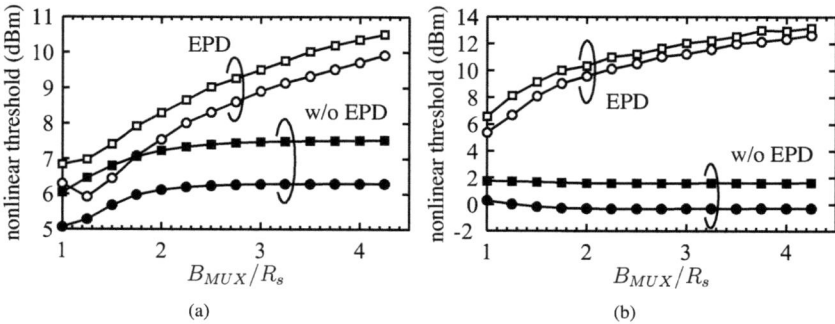

Figure 4.28: Nonlinear threshold as a function of normalised multiplexer filter bandwidth for transmission of single-channel (a) 10 Gb/s and (b) 40 Gb/s NRZ-OOK over 10×80km SSMF with target BER of 10^{-3} (squares) and 10^{-9} (circles).

and fibre type. For constant C_2, the NLT scales only with the nonlinear phase shift, such that $\phi_{NL} = $ const. For spectral efficiency 0.1 b/s/Hz and $C_1 \geq 0.8$, the NLT for WDM transmission approaches that of the single channel case. This indicates that transmission impairments begin to stem from intrachannel nonlinearities which are not completely precompensated for by EPD.

The spectral width of the predistorted signal increases with the cumulated nonlinear phase shift. For increasing launch power, the ability to precompensate for intrachannel nonlinearities thus becomes limited by the bandwidth of WDM multiplexer filters. In order to assess this limitation, simulations of single-channel transmission with varying multiplexer filter bandwidth are conducted. The transmission fibre is a SSMF. Precompensation is fixed at -232 ps/nm, which is the analytically determined optimum value to minimise intrachannel nonlinearities in this configuration[6] [59]. Fig. 4.28(a) shows the NLT as a function of multiplexer filter bandwidth B_{MUX} for transmission of 10 Gb/s single-channel NRZ-OOK. In this configuration, intra-pulse SPM is the dominant impairment. With increasing launch power, the spectrum of the predistorted signal broadens. Therefore, the ability to precompensate is severely limited for narrow multiplexer filter bandwidth. For $B_{MUX} = 1.25R_s$, application of EPD yields a NLT which is less than 1 dB higher than it would be without EPD. Fig. 4.28(b) shows the NLT for 40 Gb/s transmission. In this case, IFWM is the main source of degradation. Although the gain in NLT is also reduced, it exceeds

---
[6]This value slightly differs from the numerically obtained optimum precompensation used for Fig. 4.27

4.3 Implications on the Design of Fibre-Optic Transmission Systems

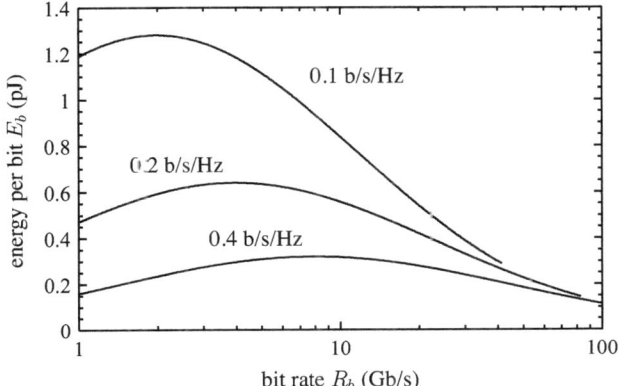

Figure 4.29: Energy per bit as a function of bit rate for transmission of electronically predistorted NRZ-OOK over 10×80 km SLAF spans with spectral efficiency 0.1, 0.2 and 0.4 b/s/Hz. The results apply to transmission of five WDM channels, i.e. the aggregate bit rate varies along each curve.

4 dB even for narrow multiplexer filter bandwidth. Depending on multiplexer filter bandwidth, bit rate and spectral efficiency, there are configurations where residual intrachannel nonlinearities due to incomplete precompensation dominate over interchannel nonlinearities. Generally, this effect will be stronger for large bit rate per channel, small spectral efficiency and narrow multiplexer bandwidth. However, the NLT of most practical configurations is rather limited by interchannel nonlinearities as suggested in [203, 204]. Thus, the parameter C_2 is a valuable tool in designing such systems and predicting their performance.

The message of Fig. 4.27(b) with respect to the NLT is quite simple. In order to maximise the NLT for a given spectral efficiency, GVD and bit rate per channel have to be chosen as large as possible. However, this strategy does not lead to a maximum energy per bit and thus OSNR margin. Based on the quadratic fit depicted in Fig. 4.27(b), a lower bound[7] for the energy per bit for transmission over 10×80 km SLAF spans (fibre parameters according to Tab. 4.3) is calculated. SLAF is chosen here because of all fibre types it enables maximum NLT for any constant C_2, when systems with EPD are considered. Fig. 4.29 shows the resulting energy per bit for spectral efficiency 0.1, 0.2 and 0.4 b/s/Hz. For small bit rates, the energy per bit becomes limited by increasing penalties due to FWM. The bit rate resulting

---

[7]Since the dispersion map is not optimised, there is potential to further increase the NLT (cp. Fig. 4.16).

4 Application to the Design of Fibre-Optic Transmission Systems

in the maximum energy per bit increases with spectral efficiency, but the maximum energy per bit is reduced. For bit rates larger than the optimum, the NLT does not increase fast enough to keep pace with the higher bit rates. Initially, the NLT increases rapidly because phase-matching and thus FWM efficiency decreases. For low FWM efficiency, XPM becomes the dominating nonlinear effect. XPM is reduced due to the walk-off between WDM channels, which leads to an averaging of the incurred nonlinear phase shifts. The smaller the walk-off length L_w according to Eq. (2.31) becomes, the more effective is the averaging operation. However, for $L_w \ll L_{eff}$, the benefit of additional averaging decreases. For this reason, the NLT shows a saturating behaviour for large C_2. Consequently, the energy per bit is reduced. Due to the saturating behaviour of the NLT the differences in energy per bit are smaller for high bit rates. This example shows the versatility and simplicity of the equivalent single-span model for the analysis and optimisation of fibre-optic communication systems.

5 Conclusion

This chapter summarises the main results of this work and discusses their application areas. Section 5.1 starts by presenting the most important parameters defining the nonlinear properties of a dispersion-managed transmission system and highlighting their significance. Thereafter, main results of exemplary numerical simulations conducted during the work on this thesis are recapitulated. This chapter concludes with a discussion of expedient extensions to the proposed equivalent single-span model.

5.1 Summary of Key Results

In this work, the concept of an equivalent single-span system – briefly touched previously in the work by Hadrien Louchet [22] – is investigated in greater detail. The mathematical tool for the derivation of the equivalent single-span model is the frequency-domain Volterra series expansion. It is reviewed in the context of the evaluation of dispersion maps in chapter 3. Starting from the Volterra transfer function of concatenated fibre sections, an approximate solution for the received electric field in single-periodic dispersion-managed transmission systems with dispersion precompensation and constant or randomly varying residual dispersion per span is derived. The influence of physical parameters of the transmission line on nonlinear propagation of the electric field is described by the nonlinear transfer function, which links the input electric field to the generated nonlinear perturbation. Analysis of the nonlinear transfer function of single-periodic dispersion-managed transmission systems reveals that there exist three parameters defining the nonlinear perturbation. These are

5 Conclusion

1. the cumulated nonlinear phase shift

$$\phi_{NL} = N\frac{\gamma P_0}{\alpha},$$

where N is the number of fibre spans, γ the fibre's nonlinear coefficient, α its attenuation coefficient and P_0 the launch power per WDM channel;

2. the 3 dB bandwidth of the nonlinear transfer function

$$\Omega_s = \frac{\alpha}{|\beta_2|},$$

where β_2 is the fibre's group-velocity dispersion;

3. the equivalent precompensation for constant residual dispersion per span

$$C'_{pre} = C_{pre} + \frac{N-1}{2}C_{res},$$

where C_{pre} is the amount of dispersion precompensation at the transmitter side and C_{res} the residual dispersion per span. For random residual dispersion per span the equivalent precompensation is

$$C'_{pre} = \frac{1}{N}\sum_{n=1}^{N} C_n,$$

where C_n is the cumulated dispersion at the beginning of the nth fibre span.

Transmitting the same input signal over different transmission lines having equal values of these parameters generates approximately the same nonlinear perturbation. Thus, all prerequisites of the model being fulfilled, there exists an equivalent single-span system for each multi-span system, leading to the same nonlinear perturbation after transmission. It is verified by numerical simulations in chapter 4 (in addition to already published results [124, 125, 165–167]) that under certain conditions, a multi-span system can be approximated by an equivalent single-span system. Dispersion precompensation of the single-span system is defined by the dispersion map parameters of the multi-span system. The single-span system is equivalent to the multi-span system in the sense that both allow the same cumulated nonlinear phase shift for a certain ROSNR penalty.

5.1 Summary of Key Results

In order to incorporate the influence of signal properties such as symbol rate and WDM channel spacing into the model, the 3-dB bandwidth and the equivalent precompensation are normalised. Normalising the 3-dB bandwidth leads to two important parameters

$$C_1 = R_s^2/\Omega_s,$$

where R_s is the symbol rate and

$$C_2 = \Delta f_{ch}^2/\Omega_s,$$

where Δf_{ch} is the WDM channel spacing.

Assuming optimised equivalent precompensation and constant spectral efficiency, either one of these parameters defines the allowed cumulated nonlinear phase shift for arbitrary fibre type and symbol rate or channel spacing, respectively. This in turn determines the maximum reasonable launch power into the system. With this information, one can readily determine the achievable OSNR at the receiver and thus the expected OSNR margin. For a certain modulation format, optimisation of a single-span system can thus be used to find optimum system parameters such as bit rate, fibre type and dispersion map for arbitrary span counts. Once established, single-span results enable rapid comparison of the nonlinear tolerance of different modulation formats and identification of their respective ideal application area.

The theoretical model is verified by numerical simulations of NRZ- and RZ-OOK transmission in section 4.2. In accordance to analytical predictions, equivalent multi-span systems in terms of ROSNR penalty exist for 40 Gb/s RZ-OOK transmission at $C_1 = 0.71$. However, this equivalence can not be found for 10 Gb/s NRZ-OOK transmission at $C_1 = 0.045$. Although optimum dispersion map parameters could still be obtained by the single-span model, the allowed cumulated nonlinear phase shift is not correctly estimated for small C_1.

Having established the theoretical framework, an exemplary analysis of single-channel and WDM RZ-OOK transmission is performed in section 4.2. It turns out that the OSNR margin for single-span WDM transmission of RZ-OCK with 33% duty cycle and 0.4 b/s/Hz spectral efficiency can be maximised by employing SLAF with a dispersion parameter $D = 20$ ps/(nm·km) and signalling with a bit rate of about 16 Gb/s per WDM channel. This configuration is optimum because it balances the effects of intrachannel and interchannel nonlinearities.

5 Conclusion

In section 4.3.1, a similar analysis is performed for RZ-DBPSK and RZ-DQPSK in order to quantify the reduction of the nonlinear threshold when upgrading from binary to quaternary modulation. Keeping the WDM channel spacing and the symbol rate constant enables a doubling of spectral efficiency from 0.4 b/s/Hz to 0.8 b/s/Hz. Furthermore, extensive numerical simulations are carried out to determine the influence of random relative phases of the optical carriers and random timings of the transmitted bit sequences in each WDM channel. Generally, the impact of these variations is strong for small C_1 but steadily decays with increasing C_1. Similar to RZ-OOK transmission, the energy per bit is maximised by employing SLAF and signalling with symbol rates of 21 GBd and 24 GBd for RZ-DBPSK and RZ-DQPSK, respectively. Doubling spectral efficiency in this way reduces the OSNR margin by about 7 dB, which makes this upgrade only feasible in systems operating with ample margin to spare. However, it should be kept in mind that RZ-DQPSK offers much better tolerance to linear impairments such as residual dispersion and polarisation-mode dispersion, such that less margin is needed to safeguard against these effects.

The influence of varying spectral efficiency on the nonlinear threshold is investigated in section 4.3.2. This investigation reveals a dualistic nature of the two parameters C_1 and C_2. On the one hand, C_1 defines the allowed cumulated nonlinear phase shift for arbitrary spectral efficiency and pseudo-linear transmission, i.e. for system configurations where the nonlinear perturbation is mainly caused by intrachannel nonlinearities. On the other hand, the same is true for C_2 but for system configurations where interchannel nonlinearities dominate. Both parameters thus describe one transmission regime successfully, while they fail in the respective other regime as soon as different spectral efficiencies are considered. This is due to C_1 and C_2 universally describing IFWM and FWM efficiency, respectively (cp. section 3.3). A general result stemming from this analysis is that the optimum value of either C_1 or C_2 resulting in maximum nonlinear threshold increases with spectral efficiency.

Compensation of fibre nonlinearity by reverse propagation models in the electronic domain has emerged as an enabling technology for future high-capacity transmission systems [201, 205, 210–212]. However, in optical networks where WDM channels are added and dropped at network nodes, it is generally impossible to compensate for interchannel nonlinearities, because phase, timing and polarisation of copropagating WDM channels are not known and can not be controlled effectively. In such systems, compensation of intrachannel nonlinearities remains the only feasible option for electronic mitigation of nonlinear impairments. Complete compensation of intrachannel nonlinearities assumed, interchannel nonlinearities remain as the only source of nonlinear impairments. As verified by numerical simulations for NRZ-OOK transmission with electronic precompensation of intrachannel nonlinearities in section 4.3.2, the parameter C_2 universally defines the nonlinear threshold in this kind of systems. The nonlinear threshold increases monotonically with C_2 and shows a saturating behaviour for large C_2. For this reason, there exists an optimum bit rate resulting in maximum OSNR margin for each spectral efficiency. It is expected that the parameter C_2 describes scaling of interchannel nonlinearities for more advanced modulation formats as well. The parameter C_2 can thus be a valuable tool for the design of future high-capacity transmission systems employing electronic compensation of intrachannel nonlinearities.

In conclusion, a simplified analytical model for the nonlinear perturbation in dispersion-managed transmissions systems is derived in this work. Scaling laws for interchannel as well as intrachannel nonlinearities are discussed. Simple strategies to maximise OSNR margin for arbitrary span count are presented. Applying these strategies, optimum system configurations for OOK as well as DPSK transmission are identified. Optimum system configurations in terms of OSNR margin are generally found in a trade-off between interchannel and intrachannel nonlinearities and by using SLAF in conjunction with an appropriate symbol rate of the transmitted signal. Furthermore, the scaling of nonlinear threshold in systems employing electronic precompensation of intrachannel nonlinearities is investigated. It was shown that the nonlinear threshold increases monotonically with the parameter C_2, which universally describes the nonlinear threshold in such systems.

5.2 Further Work or Shortcomings and Pitfalls

The limited validity of the equivalent single-span model in the presence of RDPS and strong interchannel nonlinearities is a somewhat disappointing situation. In order to better approximate the impact of RDPS on the NLTF's magnitude, there have been attempts to introduce not only an equivalent precompensation but also an equivalent 3-dB bandwidth, which then also depends on the RDPS. However, this approach necessitates a correction to the phase term of the NLTF, which would destroy its simplicity. Unfortunately, these attempts were not successful and the problem of improving the accuracy for such system configurations has still to be solved.

5 Conclusion

Due to the nature of its derivation the equivalent single-span model does not capture all detrimental nonlinear effects. Most notably in this respect is the nonlinear interaction between signal and noise during propagation, leading to nonlinear phase noise. This nonlinear interaction can become the most limiting nonlinear effect in transmission of phase-modulated signals over certain transmission lines[1]. A simple estimation of the impact of nonlinear phase noise showed that it dominates over linear phase noise for a nonlinear phase shift $\Phi_{NL}^2 > 3/4$. Incorporating this effect into the equivalent single-span model would therefore greatly enhance its ability to predict system performance in such systems and thus widen its area of applicability. A starting point could be a parametric gain approach to the analysis of single-channel DBPSK and DQPSK signals impaired by nonlinear phase noise presented in [193]. It is shown that nonlinear phase noise can be described by an equivalent ASE-noise model in many system configurations. This equivalent ASE-noise model is then applied in conjunction with Karhunen-Loève expansion to estimate the BER. It should in principle be possible to incorporate this equivalent ASE-noise model into the equivalent single-span model.

Another simplification made while deriving the equivalent single-span model concerns the nonlinear impact of DCFs. Although the input power into DCFs can be made small enough in order to make their nonlinear impact vanishingly small, this is not necessarily the optimum configuration. In fact, increasing the power into the DCF leads to a larger achievable OSNR at the receiver. Therefore, the power into the DCF has to be optimised in a trade-off between their nonlinear impact and achievable OSNR. Furthermore, conclusions about the optimum fibre dispersion parameter may change slightly. This is because a larger dispersion parameter necessitates longer DCFs resulting in a reduced achievable OSNR. Incorporating the nonlinear impact of DCFs into the equivalent single-span may be possible but certainly destroys the beauty of its simplicity.

An important property of the electric field only briefly mentioned in chapter 2 is its state of polarisation. Throughout this work only linearly polarised electric fields with a constant state of polarisation during propagation have been considered. This is justified as long as the signal is fully polarised at the transmitter, i.e. all power is concentrated in a single state of polarisation, and polarisation-mode dispersion can be neglected. However, techniques such as polarisation interleaving to reduce

---

[1] Nonlinear phase noise seems to be less of an issue for coherent systems, since there exist techniques to mitigate nonlinear phase noise by digital signal processing at the receiver [213].

5.2 Further Work or Shortcomings and Pitfalls

nonlinear effects or polarisation-multiplexing to increase spectral efficiency do not verify this condition. In order to extend applicability of the equivalent single-span model to such signals, the Manakov equation in the limit of vanishing polarisation-mode dispersion (see e.g. [214]) seems to be a promising starting point.

Acronyms

AM	amplitude modulation
ASE	amplified spontaneous emission
ASK	amplitude-shift keying
AWGN	additive white Gaussian noise
BER	bit-error ratio
BPF	band-pass filter
CW	continuous-wave
DBBS	de Bruijn binary sequence
DBPSK	differential binary phase-shift keying
DPSK	differential phase-shift keying
DCF	dispersion-compensating fibre
DCM	dispersion-compensating module
DMUX	demultiplexer
DSF	dispersion-shifted fibre
DQPSK	differential quadrature phase-shift keying
ECOC	European Conference on Optical Communication
EDFA	erbium-doped fibre amplifier
EPD	electronic predistortion
ETDM	electrical time-division multiplexing

FEC	forward error correction
FM	frequency modulation
FTTH	fibre to the home
FWM	four-wave mixing
GVD	group-velocity dispersion
IDF	inverse dispersion fibre
IFWM	intrachannel four-wave mixing
IP	internet protocol
ITU	International Telecommunication Union
IXPM	intrachannel cross-phase modulation
LTI	linear time-invariant
LPF	low-pass filter
MZM	Mach-Zehnder modulator
MUX	multiplexer
NLT	nonlinear threshold
NLTF	nonlinear transfer function
NRZ	non-return-to-zero
NLSE	nonlinear Schrödinger equation
NZDSF	non-zero dispersion-shifted fibre
OA	optical amplifier
OFC	Optical Fiber Communication Conference
OOK	on-off keying
OSNR	optical signal-to-noise ratio
OTDM	optical time-division multiplexing

PD	photodiode
PM	phase modulation
PRQS	pseudo-random quaternary sequence
PSK	phase-shift keying
QPSK	quadrature phase-shift keying
RDPS	residual dispersion per span
ROSNR	required optical signal-to-noise ratio
Rx	receiver
RZ	return-to-zero
SLAF	super-large-effective-area fibre
SMF	single-mode fibre
SPM	self-phase modulation
SSMF	standard single-mode fibre
TDM	time-division multiplexing
Tx	transmitter
WDM	wavelength-division multiplexing
XOR	exclusive or
XPM	cross-phase modulation

List of Symbols

A	complex envelope of the electric field ($W^{-1/2}$)
A_{eff}	effective core area (m^2)
B_{MUX}	full 3-dB bandwidth of optical multiplexer filter (Hz)
C	cumulated dispersion (s^2)
c	speed of light (m/s)
C_1	dispersion cumulated over effective length normalised to squared symbol rate
C_2	dispersion cumulated over effective length normalised to squared WDM channel spacing
C_{DCM}	cumulated dispersion of a dispersion compensation module (s^2)
C_n	cumulated dispersion at the input of the nth span (s^2)
C_{pre}	amount of dispersion precompensation (s^2)
C'_{pre}	equivalent precompensation (s^2)
C_{res}	residual dispersion per span (s^2)
C_{SMF}	cumulated dispersion of a single-mode fibre (s^2)
D	dispersion parameter (s/m^2)
D_{pre}	dispersion precompensation (s/m)
D'_{pre}	equivalent precompensation (s/m)
D_{res}	residual dispersion per span (s/m)
d_{ijk}	four-wave mixing degeneracy factor
ER	extinction ratio (dB)
f	frequency (Hz)
f_0	reference frequency (Hz)
F_{OA}	noise figure of an optical amplifier (dB)
L	fibre length (m)
L_{eff}	effective length (m)
L_D	dispersion length (m)
L_{NL}	nonlinear length (m)
L_w	walk-off length (m)
m	order of a de Bruijn binary sequence
M	length of the symbol alphabet
n	refractive index
n_l	linear part of the refractive index
n_{nl}	nonlinear perturbation of the refractive index

n_2	nonlinear index coefficient (m^2/W)
N	number of spans
N_{ch}	number of WDM channels
NLT	nonlinear threshold (W or dBm)
$P(z)$	time-averaged optical power (W)
P_0	average launch power (W)
P_{in}	average launch power (dBm)
R_b	bit rate (b/s)
R_s	symbol rate (Bd)
S	spectral efficiency (b/s/Hz)
t	time (s)
T	retarded time (s)
T_0	pulse width (s)
T_s	symbol duration (s)
U	normalised complex envelope of the electric field
v_g	group velocity (m/s)
z	spatial coordinate, usually chosen as direction of propagation (m)
α	attenuation (Np/m or dB/m)
β	linear part of the propagation constant (1/m)
β_2	group-velocity dispersion (s^2/m)
β'	propagation constant (1/m)
ΔC_{res}	deviation of th residual dispersion per span from its nominal value (s^2)
Δf	frequency separation (Hz)
Δf_{ch}	WDM channel spacing (Hz)
$\Delta \lambda$	noise bandwidth (m)
$\Delta \tau$	group delay difference (s)
$\Delta \Omega$	angular frequency mismatch (rad^2/s^2)
$\Delta \omega$	angular frequency difference (rad/s)
$\Delta \omega_{ch}$	WDM channel spacing (rad/s)
$\Delta \omega_s$	signal bandwidth (rad/s)
γ	nonlinear coefficient (W^{-1}m^{-1})
τ	normalised time
Ω	normalised angular frequency
ω_0	reference angular frequency (rad/s)
ζ	normalised direction of propagation

List of Figures

1.1 Spectrum of five WDM channels, each carrying information with bit rate R_b. Spectral efficiency is increased by reducing the channel spacing Δf_{ch}. 2

1.2 Selected experimental results reported in post-deadline sessions of recent OFCs and ECOCs. (a) Product of total capacity and distance versus year of presentation, (b) product of total capacity and distance versus achieved spectral efficiency. White symbols represent experiments using direct detection. Black symbols denote experiments with coherent detection where a digital sampling oscilloscope with subsequent off-line processing of the received data has been used. . . 3

2.1 Schematic of a fibre-optic transmission system consisting of a transmitter (Tx), a receiver (Rx) and N spans. Each span comprises of a single-mode fibre (SMF), a dispersion-compensating fibre (DCF) and (a) single-stage optical amplifier (OA) or (b) dual stage OA. P_0 and P_{DCF} denote output powers of the OAs in front of SMF and DCF, respectively. 6

2.2 Optical power as a function of distance for transmission lines with (a) single-stage and (b) dual-stage optical amplifiers. Grey areas indicate effective lengths of SMF and DCF. 8

2.3 Spectrogram (in logarithmic scale) of two Gaussian pulses (a) at the input of the fibre and (b) broadened by GVD at $z = 4L_D$. Insets show pulse shapes in the time domain. 17

2.4 FWM efficiency as a function of channel spacing Δf_{ch} for transmission of two wavelength channels over 80 km fibre with fibre loss $\alpha = 0.2$ dB/km and varying dispersion parameter. The curves correspond to dispersion parameters $D = 1, 4, 8, 12, 16$ and 20 ps/(nm·km) at $\lambda = 1.55$ µm. 22

3.1 Block diagram of the optical fibre modelled as a nonlinear time-invariant system. Time-domain inputs and outputs are linked through Volterra kernels $h_1, \ldots, h_N$ and frequency-domain inputs and outputs by kernel transforms $H_1, \ldots, H_N$. 25

3.2 Normalised magnitude and phase of the nonlinear transfer function $\eta_s(\Delta\Omega)$ of a single fibre with $L = 80$ km, $\alpha = 0.2$ dB/km and (a,b) $D = 4$ ps/(nm·km), i.e. $\Omega_s = 9 \cdot 10^{21}$ rad^2/s^2, and (c,d) $D = 16$ ps/(nm·km), i.e. $\Omega_s = 2.25 \cdot 10^{21}$ rad^2/s^2, at a wavelength of $\lambda = 1.55$ μm. Shown are the exact function according to Eq. (3.11) (circles) and its approximation according to Eq. (3.13) (line). 29

3.3 Fibre-optic transmission line consisting of M fibre sections and optical amplifiers. 29

3.4 (a) Cumulated gain for lumped amplification using EDFAs at positions z_i, (b) cumulated gain for distributed Raman amplification using backward pumping, (c) cumulated dispersion for SMF spans with full inline dispersion compensation using DCFs approximated as ideal lumped dispersion-compensating devices at positions z_j and (d) cumulated dispersion for SMF spans with full inline dispersion compensation using inverse dispersion fibre (IDF). 30

3.5 Block diagram of a fibre section with kernel transforms $H_{1,m}(\omega)$ and $H_{3,m}(\omega)$ preceded by an optical amplifier with transfer function $H_{a,m}(\omega)$ adding noise $\tilde{n}_m(\omega)$ to the signal. 32

3.6 Schematic of a single-span system with a dispersion precompensation stage after the transmitter. 34

3.7 Schematic of a fibre-optic transmission line consisting of N identical spans and dispersion pre- and postcompensation stages. 35

3.8 Profile of the cumulated gain $G(z)$ and cumulated dispersion $C(z)$ in a single-periodic dispersion map with dispersion precompensation C_{pre} and RDPS C_{res}. 36

3.9 Solid lines show the magnitude of the nonlinear transfer function according to Eq. (3.28). The transmission line consists of 10×80 km SSMF spans with RDPS corresponding to (a) 40 km and (b) 2 km of uncompensated SSMF. Dashed lines indicate the magnitude of $\eta_s(\Delta\Omega)$ according to Eq. (3.13). Magnitudes are normalised to their respective maximum values. 37

3.10 Probability density function (PDF) of the residual dispersion of an individual span. The uniform distribution has mean $\langle C_{res}\rangle$ and its boundaries are defined by the cumulated dispersion C_{DCM} of the employed DCM. 40

3.11 Magnitude of the nonlinear transfer function for 10×80 km SSMF and RDPS distributed with (a) $\langle C_{res}\rangle = -816$ ps^2, $C_{DCM} = 413$ ps^2 (corresponding to DCM-20) and (b) $\langle C_{res}\rangle = -41$ ps^2, $C_{DCM} = 41$ ps^2 (corresponding to DCM-2). Dashed lines indicate the magnitude of $\eta_s(\Delta\Omega)$. 42

3.12 Spectrogram representation of two dispersive Dirac pulses (bold lines) and the formation of IFWM. Pulses A and B are centred at time instants t_a and t_b and at reference angular frequency ω_0 in time and frequency domain, respectively. Their separation in the frequency domain is $2\Delta\omega$. The vertical axis represents the efficiency of the FWM process with 3-dB bandwidth Ω_s. 45

3.13 Efficiency of FWM and IFWM as functions of normalised cumulated dispersion (a) with respect to symbol rate R_s and (b) to channel spacing Δf_{ch}. The four shown cases of spectral efficiency are 0.1, 0.2, 0.4 and 0.8 b/s/Hz (binary modulation assumed). 47

4.1 Generic setup of a WDM transmission system consisting of N spans. 52
4.2 Schematic of basic OOK transmitters 52
4.3 Schematics of (a) DBPSK and (b) DQPSK transmitters and associated constellation diagrams. [141] 53
4.4 Schematics of (a) the receiver model for direct detection of intensity modulated signals, (b) receiver model for differential demodulation and balanced detection of DBPSK and (c) the same for DQPSK. . . 58
4.5 (a) Obtained and required OSNR as a function of launch power for transmission of 40 Gb/s NRZ-OOK over 10×80 km SSMF. The nonlinear threshold (NLT) is defined as the launch power resulting in 1 dB ROSNR penalty. (b) OSNR margin as a function of launch power. The NLT does not coincide with the launch power giving maximum OSNR margin. 60

4.6 Nonlinear threshold as a function of normalised precompensation for transmission of five WDM channels over a single span of fibre with (a) NRZ-OOK and spectral efficiency $S = 0.2$ b/s/Hz at normalised dispersion $C_1 = 0.0445$, (b) RZ-OOK and $S = 0.4$ b/s/Hz at $C_1 = 0.71$. 62

4.7 Nonlinear phase shift Φ_{NL} for $P_0 = \text{NLT}$ (in units Watt) as a function of normalised precompensation. The signal is transmitted over 80 km fibre with dispersion parameter $D = 16$ ps/(nm·km) and varying effective core area. (a) 5×10 Gb/s NRZ-OOK with spectral efficiency $S = 0.2$ b/s/Hz, (b) 5×40 Gb/s RZ-OOK with $S = 0.4$ b/s/Hz. 63

4.8 Nonlinear phase shift Φ_{NL} for $P_0 = \text{NLT}$ (in units Watt) as a function of normalised precompensation. The signal is transmitted over 80 km fibre with varying fibre loss. The dispersion parameter is adjusted in each case to keep C_1 constant. (a) 5×10 Gb/s NRZ-OOK with spectral efficiency $S = 0.2$ b/s/Hz, (b) 5×40 Gb/s RZ-OOK with $S = 0.4$ b/s/Hz. 64

4.9 Nonlinear threshold in dBm (contour line labels) for RZ-OOK modulation and transmission of (a) a single channel with symbol rate R_s and (b) a $5 \times R_s$ WDM signal with 0.4 b/s/Hz spectral efficiency over a single span with precompensation C_{pre}. Circles indicate configurations giving maximum nonlinear threshold for dispersion parameter $D = 4, 8, 16$ ps/(nm·km) and a bit rate of 10 Gb/s (grey dots) or 40 Gb/s (black dots). 65

4.10 Nonlinear threshold for optimised normalised precompensation $R_s^2 C_{pre}$ in case of single-channel (white circles) and $5 \times R_s$ WDM (black circles) transmission. Due to insufficient bit-sequence length, the nonlinear threshold is likely to be overestimated in the shaded area, i.e. for large channel memory. 66

4.11 OSNR margin as a function of symbol rate for transmission of $5 \times R_s$ RZ-OOK with a spectral efficiency of 0.4 b/s/Hz over 100 km fibre with optimised precompensation and (a) single-stage amplification or (b) dual-stage amplification. 68

4.12 Nonlinear phase shift Φ_{NL} for $P_0 = \text{NLT}$ (in units Watt) as a function of normalised precompensation for transmission over N spans of SSMF. (a) 10 Gb/s NRZ-OOK with spectral efficiency $S = 0.2$ b/s/Hz, (b) 40 Gb/s RZ-OOK with $S = 0.4$ b/s/Hz. 70

4.13 ROSNR penalty in dB (contour line labels) as a function of dispersion map parameters for transmission of 5×10 Gb/s NRZ-OOK over 10×80 km SSMF with 4 dBm launch power and spectral efficiency 0.2 b/s/Hz. Black dots indicate numerically obtained optimum precompensation for each value of RDPS. The solid line represents optimum precompensation as a function of RDPS obtained semi-analytically with the equivalent single-span model. Optimum precompensation of the single-span system C^*_{pre} is determined numerically [cp. Fig.4.6(a)]. Then Eq. (3.29) is used to calculate optimum precompensation for the multi-span system. Dashed line represents analytically predicted optimum precompensation as a function of RDPS according to Killey's formula [Eq. (3.55)]. 71

4.14 ROSNR penalty in dB (contour line labels) as a function of dispersion map parameters for transmission of 5×40 Gb/s RZ-OOK over 10×80 km SSMF with 2 dBm launch power and spectral efficiency 0.4 b/s/Hz. Black dots indicate numerically obtained optimum precompensation for each value of RDPS. Solid and dashed line represent predicted optimum precompensation as a function of RDPS according to the equivalent single-span model and Killey's formula [Eq. (3.55)], respectively. 73

4.15 ROSNR penalty as a function of RDPS for transmission of 5×10 Gb/s NRZ-OOK and 5×40 Gb/s RZ-OOK over 10 spans of SSMF. The equivalent precompensation is kept constant. This corresponds to the ROSNR penalty along the solid lines in Fig. 4.13 and 4.14. The grey area represents the validity region of the equivalent single-span model according to Eq. (3.36). 74

4.16 Nonlinear phase shift Φ_{NL} for $P_0 = \text{NLT}$ (in units Watt) as a function of span count for transmission over N spans of SSMF with optimised dispersion map. 75

4.17 ROSNR penalty as a function of equivalent precompensation. Shown are the results of split-step Fourier simulations for transmission of 5×40 Gb/s RZ-OOK over different transmission lines: a single 80 km span of SSMF with launch power of 12 dBm per channel (circles), and 10×80 km SSMF with uniformly distributed residual dispersion per span, varying precompensation and 2 dBm launch power per channel (dots). The distribution parameters are: (a) mean RDPS $\langle C_{res}\rangle = \pm 51$ ps^2, width $C_{DCM} = 102$ ps^2 and (b) mean RDPS $\langle C_{res}\rangle = \pm 102$ ps^2, width $C_{DCM} = 204$ ps^2. 77

4.18 (a) Spectra of RZ-DBPSK and RZ-DQPSK for equal channel spacing and symbol rate and (b) associated time-domain waveforms of four exemplary symbols. 80

4.19 NLT for single-span transmission of $5\times R_s$ RZ-DBPSK and RZ-DQPSK as a function of normalised dispersion C_1. Grey areas are bounded by the respective 10th and 90th percentiles of the NLT distribution resulting from 100 simulation runs with random timings of the transmitted bit sequences in the five WDM channels and random relative carrier phases. 81

4.20 Energy per symbol resulting in 1 dB ROSNR penalty as a function of symbol rate for transmission over a single-span of 80 km SLAF with optimised precompensation. 82

4.21 Contour plot of ROSNR penalty as a function of dispersion precompensation and RDPS for transmission of (a) 5×10 GBd with $P_{in} = 0$ dBm and (b) 5×40 GBd RZ-DQPSK with $P_{in} = 4$ dBm over 10×80 km of SSMF. Black dots indicate optimum amount of precompensation for each simulated value of RDPS. Solid line is optimum precompensation as a function of RDPS according to Eq. (3.29), where the optimum value of the equivalent precompensation C'_{pre} was numerically obtained. Dashed line is optimum precompensation according to Eq. 3.55. 84

4.22 ROSNR penalty (contour line labels) as a function of dispersion map parameters for transmission of 5×20 GBd RZ-DQPSK over 10×80 km SLAF with $P_{in} = 4$ dBm. 85

4.23 (a) Mismatch of normalised precompensation $C_{pre}R_s^2$ and (b) difference in ROSNR penalty between analytically and numerically obtained optima versus normalised RDPS. 86

4.24 ROSNR penalty as a function of RDPS for system configurations shown in Fig. 4.21 and 4.22. The equivalent precompensation is kept constant. This corresponds to the ROSNR penalty along the solid line representing optimum precompensation in those figures. The grey area represents the validity region of the equivalent single-span model according to Eq. (3.36). 88

4.25 Nonlinear threshold for single-span transmission of NRZ-OOK with optimised precompensation and varying spectral efficiency S as a function of (a) normalised dispersion C_1 and (b) normalised dispersion C_2. 89

4.26 Schematic of the system setup with EPD of the centre channel. . . . 90

4.27 Nonlinear threshold for transmission of electronically precompensated NRZ-OOK over 10×80 km spans with optimised precompensation and varying spectral efficiency S as a function of (a) normalised dispersion C_1 and (b) normalised dispersion C_2. The solid line represents a quadratic fit to the data. 91

4.28 Nonlinear threshold as a function of normalised multiplexer filter bandwidth for transmission of single-channel (a) 10 Gb/s and (b) 40 Gb/s NRZ-OOK over 10×80km SSMF with target BER of 10^{-3} (squares) and 10^{-9} (circles). 92

4.29 Energy per bit as a function of bit rate for transmission of electronically predistorted NRZ-OOK over 10×80 km SLAF spans with spectral efficiency 0.1, 0.2 and 0.4 b/s/Hz. The results apply to transmission of five WDM channels, i.e. the aggregate bit rate varies along each curve. 93

Bibliography

[1] A. Gladisch, R.-P. Braun, D. Breuer, A. Ehrhardt, H.-M. Foisel, M. Jaeger, R. Leppla, M. Schneiders, S. Vorbeck, W. Weiershausen, and F.-J. Westphal, "Evolution of terrestrial optical system and core network architecture," *Proceedings of the IEEE*, vol. 94, no. 5, pp. 869–891, May 2006.

[2] D. Gallant, "Optical network foundation for triple play services roll-out," in *Proc. Optical Fiber Communication Conference and National Fiber Optic Engineers Conference, OFC/NFOEC'06*, Anaheim, USA, March 2006, paper NWC3.

[3] A. Ehrhardt, "Next generation optical networks and new services: an operator's point of view," in *Proc. 9th International Conference on Transparent Optical Networks, ICTON'07*, Rome, Italy, July 2007, paper Tu.B4.2.

[4] T. Fujii, "Recent progress of digital cinema over optical networks," in *Proc. Optical Fiber Communication Conference and National Fiber Optic Engineers Conference, OFC/NFOEC'05*, Anaheim, USA, March 2005, paper OFP4.

[5] E. Desurvire, "Optical communications in 2025," in *Proc. 31st European Conference on Optical Communication, ECOC'05*, Glasgow, Scotland, September 2005, paper Mo2.1.3.

[6] A. M. J. Koonen, A. Ng'oma, G.-J. Rijckenberg, M. Garcia Larrode, P. J. Urban, H. de Waardt, J. Yang, H. Yang, and H. P. A. van den Boom, "How deep should fibre go into the access network?" in *Proc. 33rd European Conference on Optical Communication, ECOC'07*, Berlin, Germany, September 2007, paper 1.1.4.

[7] P. J. Winzer, "Modulation and multiplexing in optical communication systems," *IEEE Lasers and Electro-Optics Society Newsletter*, vol. 23, no. 1, pp. 4–10, February 2009.

[8] K. Schuh, E. Lach, B. Junginger, and G. Veith, "8 Tbit/s (80×107 Gbit/s) DWDM ASK-NRZ VSB transmission over 510 km NZDSF with 1bit/s/Hz spectral efficiency," in *Proc. 33rd European Conference on Optical Communication, ECOC'07*, Berlin, Germany, September 2007, paper PDS 1.8.

[9] J. G. Proakis and M. Salehi, *Digital Communications*, 5th ed. McGraw-Hill, 2008.

[10] L. E. Nelson, S. L. Woodward, S. Foo, X. Zhou, M. D. Feuer, D. Hanson, D. McGhan, H. Sun, M. Moyer, M. O. Sullivan, and P. D. Magill, "Performance of a 46-Gbps dual-polarization QPSK transceiver with real-time coherent equalization over high PMD fiber," *Journal of Lightwave Technology*, vol. 27, no. 3, pp. 158–167, February 2009.

[11] P. P. Mitra and J. B. Stark, "Nonlinear limits to the information capacity of optical fibre communications," *Nature*, vol. 411, no. 6841, pp. 1027–1030, June 2001.

[12] J. Tang, "The multispan effects of Kerr nonlinearity and amplifier noises on Shannon channel capacity of a dispersion-free nonlinear optical fiber," *Journal of Lightwave Technology*, vol. 19, no. 8, pp. 1110–1115, August 2001.

[13] A. Mecozzi and M. Shtaif, "On the capacity of intensity modulated systems using optical amplifiers," *IEEE Photonics Technology Letters*, vol. 13, no. 9, pp. 1029–1031, September 2001.

[14] E. E. Narimanov and P. Mitra, "The channel capacity of a fiber optics communication system: perturbation theory," *Journal of Lightwave Technology*, vol. 20, no. 3, pp. 530–537, March 2002.

[15] J. Tang, "The channel capacity of a multispan DWDM system employing dispersive nonlinear optical fibers and an ideal coherent optical receiver," *Journal of Lightwave Technology*, vol. 20, no. 7, pp. 1095–1101, July 2002.

[16] E. Desurvire, "Quantum noise model for ultimate information-capacity limits in long-haul WDM transmission," *Electronics Letters*, vol. 38, no. 17, pp. 983–984, August 2002.

[17] K. S. Turitsyn, S. A. Derevyanko, I. V. Yurkevich, and S. K. Turitsyn, "Information capacity of optical fiber channels with zero average dispersion," *Physical Review Letters*, vol. 91, no. 20, p. 203901, November 2003.

[18] L. G. L. Wegener, M. L. Povinelli, A. G. Green, P. P. Mitra, J. B. Stark, and P. B. Littlewood, "The effect of propagation nonlinearities on the information capacity of WDM optical fiber systems: cross-phase modulation and four-wave mixing," *Physica D: Nonlinear Phenomena*, vol. 189, no. 2, pp. 81–99, February 2004.

[19] J. M. Kahn and K.-P. Ho, "Spectral efficiency limits and modulation/detection techniques for DWDM systems," *IEEE Journal of Selected Topics in Quantum Electronics*, vol. 10, no. 2, pp. 259–272, March/April 2004.

[20] J. Tang, "A comparison study of the Shannon channel capacity of various nonlinear optical fibers," *Journal of Lightwave Technology*, vol. 24, no. 5, pp. 2070–2075, May 2006.

[21] R.-J. Essiambre, G. J. Foschini, G. Kramer, and P. J. Winzer, "Capacity limits of information transport in fiber-optic networks," *Physical Review Letters*, vol. 101, no. 16, p. 163901, October 2008.

[22] H. Louchet, "Top-down analysis of high-capacity fiber-optic transmission," Ph.D. dissertation, Technische Universität Berlin, Berlin, Germany, October 2006.

[23] G. P. Agrawal, *Nonlinear Fiber Optics*, 4th ed. Academic Press, December 2006.

[24] ——, *Fiber-Optic Communication Systems*, 3rd ed. Wiley & Sons, 2002, ch. 2.5 Fiber Loss, pp. 55–59.

[25] T. Miya, Y. Terunuma, T. Hosaka, and T. Miyashita, "Ultimate low-loss single-mode fibre at 1.55 μm," *Electronics Letters*, vol. 15, no. 4, pp. 106–108, February 1979.

[26] K. Nagayama, M. Kakui, M. Matsui, I. Saitoh, and Y. Chigusa, "Ultra-low-loss (0.1484 dB/km) pure silica core fibre and extension of transmission distance," *Electronics Letters*, vol. 38, no. 20, pp. 1168–1169, September 2002.

[27] E. Desurvire, *Erbium-doped fiber amplifiers – Principles and applications*. John Wiley & Sons, 1994.

[28] G. P. Agrawal, *Fiber-Optic Communication Systems*, 3rd ed. Wiley & Sons, 2002, ch. 6 Optical Amplifiers, pp. 226–278.

[29] R.-J. Essiambre, G. Raybon, and B. Mikkelsen, *Optical Fiber Telecommunications IVB*. Academic Press, 2002, ch. 6 Pseudo-Linear Transmission of High-Speed TDM Signals: 40 and 160 Gb/s.

[30] P. J. Winzer and R.-J. Essiambre, "Advanced optical modulation formats," *Proceedings of the IEEE*, vol. 9, no. 5, pp. 952–985, May 2006.

[31] R. H. Stolen and A. Ashkin, "Optical Kerr effect in glass waveguide," *Applied Physics Letters*, vol. 22, no. 6, pp. 294–296, March 1973.

[32] K. S. Kim, R. H. Stolen, W. A. Reed, and K. W. Quoi, "Measurement of the nonlinear index of silica-core and dispersion-shifted fibers," *Optics Letters*, vol. 19, no. 4, pp. 257–259, February 1994.

[33] J.-C. Antona, S. Bigo, and S. Kosmalski, "Nonlinear index measurements of various fibre types over C+L bands using four-wave mixing," in *Proc. 27th European Conference on Optical Communication, ECOC'01*, Amsterdam, The Netherlands, September 2001, paper We.L.1.2.

[34] M. Bigot-Astruc, F. Gooijer, N. Montaigne, and P. Sillard, "Trench-assisted profiles for large-effective-area single-mode fibers," in *Proc. 34th European Conference on Optical Communication, ECOC'08*, Brussels, Belgium, September 2008, paper Mo.4.B.1.

[35] G. Charlet, M. Salsi, H. Mardoyan, P. Tran, J. Renaudier, S. Bigo, M. Astruc, P. Sillard, L. Provost, and F. Cérou, "Transmission of 81 channels at 40Gbit/s over a transpacific-distance erbium-only link, using PDM-BPSK modulation, coherent detection and a new large effective area fibre," in *Proc. 34th European Conference on Optical Communication, ECOC'08*, Brussels, Belgium, September 2008, paper Th.3.E.3.

[36] R. Ludwig, W. Pieper, H. G. Weber, D. Breuer, K. Petermann, F. Küppers, and A. Mattheus, "Unrepeatered 40 Gbit/s RZ single channel transmission over 150 km of standard singlemode fibre at 1.55 µm," *Electronics Letters*, vol. 33, no. 1, pp. 76–77, January 1997.

[37] D. Breuer and K. Petermann, "Comparison of NRZ- and RZ-modulation format for 40-Gb/s TDM standard-fiber systems," *IEEE Photonics Technology Letters*, vol. 9, no. 3, pp. 398–400, March 1997.

[38] K.-I. Suzuki, N. Ohkawa, M. Murakami, and K. Aida, "Unrepeatered 40 Gbit/s RZ signal transmission over 240 km conventional singlemode fibre," *Electronics Letters*, vol. 34, no. 8, pp. 799–800, April 1998.

[39] K. Ennser, R. I. Laming, and M. N. Zervas, "Analysis of 40 Gb/s TDM-transmission over embedded standard fiber employing chirped fiber grating dispersion compensators," *Journal of Lightwave Technology*, vol. 16, no. 5, pp. 807–811, May 1998.

[40] D. Breuer, H. J. Ehrke, F. Küppers, R. Ludwig, K. Petermann, H. G. Weber, and K. Weich, "Unrepeated 40-Gb/s RZ single-channel transmission at 1.55 μm using various fiber types," *IEEE Photonics Technology Letters*, vol. 10, no. 6, pp. 822–824, June 1998.

[41] R.-J. Essiambre, B. Mikkelsen, and G. Raybon, "Intra-channel cross-phase modulation and four-wave mixing in high-speed TDM systems," *Electronics Letters*, vol. 35, no. 18, pp. 1576–1578, September 1999.

[42] C. M. Weinert, R. Ludwig, W. Pieper, H. G.Weber, D. Breuer, K. Petermann, and F. Küppers, "40 Gb/s and 4×40 Gb/s TDM/WDM standard fiber transmission," *Journal of Lightwave Technology*, vol. 17, no. 11, pp. 2276–2284, November 1999.

[43] Y. Miyamoto, A. Hirano, K. Yonenaga, A. Sano, H. Toba, K. Murata, and O. Mitomi, "320 Gbit/s (8×40 Gbit/s) WDM transmission over 367 km with 120 km repeater spacing using carrier-suppressed return-to-zero format," *Electronics Letters*, vol. 35, no. 23, pp. 2041–2042, November 1999.

[44] R. Ludwig, U. Feiste, E. Dietrich, H. G. Weber, D. Breuer, M. Martin, and F. Küppers, "Experimental comparison of 40 Gbit/s RZ and NRZ transmission over standard singlemode fibre," *Electronics Letters*, vol. 35, no. 25, pp. 2216–2218, December 1999.

[45] A. H. Gnauck, S.-G. Park, J. M. Wiesenfeld, and L. D. Garrett, "Highly dispersed pulses for 2×40 Gbit/s transmission over 800 km of conventional singlemode fibre," *Electronics Letters*, vol. 35, no. 25, pp. 2218–2219, December 1999.

[46] L. K. Wickham, R.-J. Essiambre, A. H. Gnauck, P. J. Winzer, and A. R. Chraplyvy, "Bit pattern length dependence of intrachannel nonlinearities in

pseudolinear transmission," *IEEE Photonics Technology Letters*, vol. 16, no. 6, pp. 1591–1593, June 2004.

[47] D. Fonseca and A. Cartaxo, "On the transition to pseudo-linear regime in dispersion managed systems with NRZ, RZ and duobinary signal formats," *IEE Proceedings - Optoelectronics*, vol. 152, no. 3, pp. 181–187, June 2005.

[48] Y. Frignac and S. Bigo, "Numerical optimization of residual dispersion in dispersion-managed systems at 40 Gbit/s," in *Proc. Optical Fiber Communication Conference, OFC'00*, Baltimore, USA, March 2000, paper TuD3.

[49] G. Bellotti, A. Bertaina, and S. Bigo, "Dependence of self-phase modulation impairments on residual dispersion in 10-Gb/s-based terrestrial transmissions using standard fiber," *IEEE Photonics Technology Letters*, vol. 11, no. 7, pp. 824–826, July 1999.

[50] R. H. Stolen and C. Lin, "Self-phase-modulation in silica optical fibers," *Physical Review A*, vol. 17, no. 4, pp. 1448–1453, April 1978.

[51] A. R. Chraplyvy, R. W. Tkach, L. L. Buhl, and R. C. Alferness, "Phase modulation to amplitude modulation conversion of CW laser light in optical fibres," *Electronics Letters*, vol. 22, no. 8, pp. 409–411, April 1986.

[52] J. Wang and K. Petermann, "Small signal analysis for dispersive optical fiber communication systems," *Journal of Lightwave Technology*, vol. 10, no. 1, pp. 96–100, January 1992.

[53] P. V. Mamyshev and N. A. Mamysheva, "Pulse-overlapped dispersion-managed data transmission and intrachannel four-wave mixing," *Optics Letters*, vol. 24, no. 21, pp. 1454–1456, November 1999.

[54] A. Mecozzi, C. Clausen, and M. Shtaif, "Analysis of intrachannel nonlinear effects in highly dispersed optical pulse transmission," *IEEE Photonics Technology Letters*, vol. 12, no. 4, pp. 392–394, April 2000.

[55] J. Mårtensson, A. Berntson, M. Westlund, A. Danielsson, P. Johannisson, D. Anderson, and M. Lisak, "Timing jitter owing to intrachannel pulse interactions in dispersion-managed transmission systems," *Optics Letters*, vol. 26, no. 2, pp. 55–57, January 2001.

[56] A. Mecozzi, C. B. Clausen, and M. Shtaif, "System impact of intra-channel nonlinear effects in highly dispersed optical pulse transmission," *IEEE Photonics Technology Letters*, vol. 12, no. 12, pp. 1633–1635, December 2000.

[57] M. J. Ablowitz and T. Hirooka, "Intrachannel pulse interactions in dispersion-managed transmission systems: timing shifts," *Optics Letters*, vol. 26, no. 23, pp. 1846–1848, December 2001.

[58] T. Hirooka and M. J. Ablowitz, "Analysis of timing and amplitude jitter due to intrachannel dispersion-managed pulse interactions," *IEEE Photonics Technology Letters*, vol. 14, no. 5, pp. 633–635, May 2002.

[59] R. I. Killey, H. J. Thiele, V. Mikhailov, and P. Bayvel, "Reduction of intrachannel nonlinear distortion in 40-Gb/s-based WDM transmission over standard fiber," *IEEE Photonics Technology Letters*, vol. 12, no. 12, pp. 1624–1626, December 2000.

[60] V. Mikhailov, R. I. Killey, S. Appathurai, and P. Bayvel, "Investigation of intra-channel nonlinear distortion in 40 Gbit/s transmission over standard fibre," in *Proc. 27th European Conference on Optical Communication, ECOC'01*, Amsterdam, The Netherlands, September 2001, paper Mo.L.3.4.

[61] W. Pieper, C. Kurtzke, R. Schnabel, D. Breuer, R. Ludwig, K. Petermann, and H. G. Weber, "Nonlinearity-insensitive standard-fibre transmission based on optical-phase conjugation in a semiconductor-laser amplifier," *Electronics Letters*, vol. 30, no. 9, pp. 724–726, April 1994.

[62] A. Chowdhury, G. Raybon, R.-J. Essiambre, J. Sinsky, A. Adamiecki, J. Leuthold, C. Doerr, and S. Chandrasekhar, "Compensation of intrachannel nonlinearities in 40-Gb/s pseudolinear systems using optical-phase conjugation," *Journal of Lightwave Technology*, vol. 23, no. 1, pp. 172–177, January 2005.

[63] I. Shake, H. Takara, K. Mori, S. Kawanishi, and Y. Yamabayashi, "Influence of inter-bit four-wave mixing in optical TDM transmission," *Electronics Letters*, vol. 34, no. 16, pp. 1600–1601, August 1998.

[64] M. J. Ablowitz and T. Hirooka, "Resonant nonlinear intrachannel interactions in strongly dispersion-managed transmission systems," *Optics Letters*, vol. 25, no. 24, pp. 1750–1752, December 2000.

[65] S. Kumar, "Intrachannel four-wave mixing in dispersion managed RZ systems," *IEEE Photonics Technology Letters*, vol. 13, no. 8, pp. 800–802, August 2001.

[66] M. J. Ablowitz and T. Hirooka, "Intrachannel pulse interactions in dispersion-managed transmission systems: energy transfer," *Optics Letters*, vol. 27, no. 3, pp. 203–205, February 2002.

[67] S. Kumar, J. Mauro, S. Raghavan, and D. Chowdhury, "Intrachannel nonlinear penalties in dispersion-managed transmission systems," *IEEE Journal of Selected Topics in Quantum Electronics*, vol. 8, no. 3, pp. 626–631, May/June 2002.

[68] F. Hlawatsch and G. F. Boudreaux-Bartels, "Linear and quadratic time-frequency signal representations," *IEEE Signal Processing Magazine*, vol. 9, no. 2, pp. 21–67, April 1992.

[69] P. Johannisson, D. Anderson, A. Berntson, and J. Mårtensson, "Generation and dynamics of ghost pulses in strongly dispersion-managed fiber-optic communication systems," *Optics Letters*, vol. 26, no. 16, pp. 1227–1229, August 2001.

[70] G. L. Meur and E. Corbel, "Benefits of bit-to-bit polarisation interleaving for N×40 Gbit/s all-distributed Raman amplified submarine transmission," *Electronics Letters*, vol. 38, no. 20, pp. 1191–1193, September 2002.

[71] A. Hodžić, B. Konrad, and K. Petermann, "Improvement of system performance in N×40-Gb/s WDM transmission using alternate polarizations," *IEEE Photonics Technology Letters*, vol. 15, no. 1, pp. 153–155, January 2003.

[72] C. Xie, I. Kang, A. H. Gnauck, L. Möller, L. F. Mollenauer, and A. R. Grant, "Suppression of intrachannel nonlinear effects with alternate-polarization formats," *Journal of Lightwave Technology*, vol. 22, no. 3, pp. 806–812, March 2004.

[73] X. Liu, C. Xu, and X. Wei, "Performance analysis of time-polarization multiplexed 40-Gb/s RZ-DPSK DWDM transmission," *IEEE Photonics Technology Letters*, vol. 16, no. 1, pp. 302–304, January 2004.

[74] A. H. Gnauck, J. Leuthold, C. Xie, I. Kang, S. Chandrasekhar, P. Bernasconi, C. Doerr, L. Buhl, J. D. Bull, N. A. F. Jaeger, H. Kato, and A. Guest, "6×42.7-Gb/s transmission over ten 200-km EDFA-amplified SSMF spans using polarization-alternating RZ-DPSK," in *Proc. Optical Fiber Communication Conference, OFC'04*, Los Angeles, USA, February 2004, paper PDP35.

[75] I. Kang, C. Xie, C. Dorrer, and A. Gnauck, "Implementations of alternate-polarisation differential-phase-shift-keying transmission," *Electronics Letters*, vol. 40, no. 5, pp. 333– 335, March 2004.

[76] G. Charlet, R. Dischler, A. Klekamp, P. Tran, H. Mardoyan, L. Pierre, W. Idler, and S. Bigo, "WDM bit-to-bit alternate-polarisation RZ-DPSK transmission at 40×42.7 Gb/s over transpacific distance with large Q-factor margin," in *Proc. 30th European Conference on Optical Communication, ECOC'04*, Stockholm, Sweden, September 2004, paper Th4.4.5.

[77] A. Klekamp, R. Dischler, and W. Idler, "Impairments of bit-to-bit alternate-polarization on non-linear threshold, CD and DGD tolerance of 43 Gb/s ASK and DPSK formats," in *Proc. Optical Fiber Communication Conference and National Fiber Optic Engineers Conference, OFC/NFOEC'05*, Anaheim, USA, March 2005, paper OFN3.

[78] G. Charlet, H. Mardoyan, P. Tran, A. Klekamp, M. Astruc, M. Lefrancois, and S. Bigo, "Upgrade of 10 Gbit/s ultra-long-haul system to 40 Gbit/s with APol RZ-DPSK modulation format," *Electronics Letters*, vol. 41, no. 22, pp. 1240–1241, October 2005.

[79] X. Liu, X. Wei, A. H. Gnauck, C. Xu, and L. K. Wickham, "Suppression of intrachannel four-wave-mixing induced ghost pulses in high-speed transmissions by phase inversion between adjacent marker blocks," *Optics Letters*, vol. 27, no. 13, pp. 1177–1179, July 2002.

[80] M. Forzati, J. Mårtensson, A. Berntson, A. Djupsjöbacka, and P. Johannisson, "Reduction of intrachannel four-wave mixing using the alternate-phase RZ modulation format," *IEEE Photonics Technology Letters*, vol. 14, no. 9, pp. 1285–1287, September 2002.

[81] S. Randel, B. Konrad, A. Hodžić, and K. Petermann, "Influence of bitwise phase changes on the performance of 160 Gbit/s transmission systems,"

in *Proc. 28th European Conference on Optical Communication, ECOC'02.* Copenhagen, Denmark, September 2002, paper P3.31.

[82] D. M. Gill, A. H. Gnauck, X. Liu, X. Wei, and Y. Su, "$\pi/2$ alternate-phase on-off keyed 42.7 Gb/s long-haul transmission over 1980 km of standard single-mode fiber," *IEEE Photonics Technology Letters*, vol. 16, no. 3, pp. 906–908, March 2004.

[83] N. Alić and Y. Fainman, "Data-dependent phase coding for suppression of ghost pulses in optical fibers," *IEEE Photonics Technology Letters*, vol. 16, no. 4, pp. 1212–1214, April 2004.

[84] M. Forzati, A. Berntson, and J. Mårtensson, "IFWM suppression using APRZ with optimized phase-modulation parameters," *IEEE Photonics Technology Letters*, vol. 16, no. 10, pp. 2368–2370, October 2004.

[85] Y. Su, L. Möller, R. Ryf, C. Xie, and X. Liu, "A 160-Gb/s group-alternating-phase CSRZ format," *IEEE Photonics Technology Letters*, vol. 17, no. 10, pp. 2233–2235, October 2005.

[86] S. Appathurai, V. Mikhailov, R. I. Killey, and P. Bayvel, "Effective suppression of intra-channel nonlinear distortion in 40 Gbit/s transmission over standard singlemode fibre using alternate-phase RZ and alternate polarisation," *Electronics Letters*, vol. 40, no. 14, pp. 897–898, July 2004.

[87] J. K. Fischer and K. Petermann, "Performance analysis of CSRZ-OOK with pairwise or pulse-to-pulse alternate polarization," *IEEE Photonics Technology Letters*, vol. 19, no. 24, pp. 1997–1999, December 2007.

[88] F. Forghieri, R. W. Tkach, and A. R. Chraplyvy, *Optical Fiber Telecommunications III A*. Academic Press, 1997, ch. 8 Fiber nonlinearities and their impact on transmission systems, pp. 196–264.

[89] P. Bayvel and R. I. Killey, *Optical Fiber Telecommunications IVB*. Academic Press, 2002, ch. 13 Nonlinear optical effects in WDM transmission, pp. 611–641.

[90] G. P. Agrawal, P. L. Baldeck, and R. R. Alfano, "Temporal and spectral effects of cross-phase modulation on copropagating ultrashort pulses in optical fibers," *Physical Review A*, vol. 40, no. 9, pp. 5063–5072, November 1989.

[91] A. Chraplyvy, D. Marcuse, and P. Henry, "Carrier-induced phase noise in angle-modulated optical-fiber systems," *Journal of Lightwave Technology*, vol. 2, no. 1, pp. 6–10, February 1984.

[92] A. R. Chraplyvy and J. Stone, "Measurement of crossphase modulation in coherent wavelength-division multiplexing using injection lasers," *Electronics Letters*, vol. 20, no. 24, pp. 996–997, November 1984.

[93] M. Shtaif and M. Eiselt, "Analysis of intensity interference caused by cross-phase modulation in dispersive optical fibers," *IEEE Photonics Technology Letters*, vol. 10, no. 7, pp. 979–981, July 1998.

[94] M. Shtaif, "Analytical description of cross-phase modulation in dispersive optical fibers," *Optics Letters*, vol. 23, no. 15, pp. 1191–1193, August 1998.

[95] D. Marcuse, A. R. Chraplyvy, and R. W. Tkach, "Dependence of cross-phase modulation on channel number in fiber WDM systems," *Journal of Lightwave Technology*, vol. 12, no. 5, pp. 885–890, May 1994.

[96] T.-K. Chiang, N. Kagi, T. K. Fong, M. E. Marhic, and L. G. Kazovsky, "Cross-phase modulation in dispersive fibers: theoretical and experimental investigation of the impact of modulation frequency," *IEEE Photonics Technology Letters*, vol. 6, no. 6, pp. 733–736, June 1994.

[97] S. Bigo, G. Bellotti, and M. W. Chbat, "Investigation of cross-phase modulation limitation over various types of fiber infrastructures," *IEEE Photonics Technology Letters*, vol. 11, no. 5, pp. 605–607, May 1999.

[98] N. Kikuchi, K. Sekine, and S. Sasaki, "Analysis of cross-phase modulation (XPM) effect on WDM transmission performance," *Electronics Letters*, vol. 33, no. 8, pp. 653–654, April 1997.

[99] R. H. Stolen, J. E. Bjorkholm, and A. Ashkin, "Phase-matched three-wave mixing in silica fiber optical waveguides," *Applied Physics Letters*, vol. 24, no. 7, pp. 308–310, April 1974.

[100] K. O. Hill, D. C. Johnson, B. S. Kawasaki, and R. I. MacDonald, "cw three-wave mixing in single-mode optical fibers," *Journal of Applied Physics*, vol. 49, no. 10, pp. 5098–5106, October 1978.

[101] ITU-T, "G.694.1 Spectral grids for WDM applications: DWDM frequency grid."

[102] ——, "G.694.2 Spectral grids for WDM applications: CWDM wavelength grid."

[103] A. R. Chraplyvy, "Limitations on lightwave communications imposed by optical-fiber nonlinearities," *Journal of Lightwave Technology*, vol. 8, no. 10, pp. 1548–1557, October 1990.

[104] N. Shibata, R. Braun, and R. Waarts, "Phase-mismatch dependence of efficiency of wave generation through four-wave mixing in a single-mode optical fiber," *IEEE Journal of Quantum Electronics*, vol. 23, no. 7, pp. 1205–1210, July 1987.

[105] R. G. Waarts and R.-P. Braun, "System limitations due to four-wave mixing in single-mode optical fibres," *Electronics Letters*, vol. 22, no. 16, pp. 873–875, July 1986.

[106] F. Forghieri, R. W. Tkach, A. R. Chraplyvy, and D. Marcuse, "Reduction of four-wave mixing crosstalk in WDM systems using unequally spaced channels," *IEEE Photonics Technology Letters*, vol. 6, no. 6, pp. 754–756, June 1994.

[107] K. Nakajima, M. Ohashi, K. Shiraki, T. Horiguchi, K. Kurokawa, and Y. Miyajima, "Four-wave mixing suppression effect of dispersion distributed fibers," , *Journal of Lightwave Technology*, vol. 17, no. 10, pp. 1814–1822, October 1999.

[108] K. V. Peddanarappagari and M. Brandt-Pearce, "Volterra series transfer function of single-mode fibers," *Journal of Lightwave Technology*, vol. 15, no. 12, pp. 2232–2241, December 1997.

[109] ——, "Volterra series approach for optimizing fiber-optic communications system designs," *Journal of Lightwave Technology*, vol 16, no. 11, pp. 2046–2055, November 1998.

[110] B. Xu and M. Brandt-Pearce, "Comparison of FWM- and XPM-induced crosstalk using the Volterra series transfer function method," *Journal of Lightwave Technology*, vol. 21, no. 1, pp. 40–53, January 2003.

[111] X. Wei, "Power-weighted dispersion distribution function for characterizing nonlinear properties of long-haul optical transmission links," *Optics Letters*, vol. 31, no. 17, pp. 2544–2546, September 2006.

[112] B. Xu and M. Brandt-Pearce, "Modified Volterra series transfer function method," *IEEE Photonics Technology Letters*, vol. 14, no. 1, pp. 47–49, January 2002.

[113] A. Vannucci, P. Serena, and A. Bononi, "The RP method: a new tool for the iterative solution of the nonlinear Schrödinger equation," *Journal of Lightwave Technology*, vol. 20, no. 7, pp. 1102–1112, July 2002.

[114] H. Louchet, A. Hodžić, K. Petermann, A. Robinson, and R. Epworth, "Simple criterion for the characterization of nonlinear impairments in dispersion-managed optical transmission systems," *IEEE Photonics Technology Letters*, vol. 17, no. 10, pp. 2089–2091, October 2005.

[115] J. K. Fischer, C.-A. Bunge, and K. Petermann, "Equivalent single-span model for dispersion-managed fibre-optic transmission systems," *Journal of Lightwave Technology*, accepted for publication.

[116] M. Schetzen, *The Volterra & Wiener theories of nonlinear systems*. Malabar, USA: Krieger Publishing Company, 2006.

[117] S. K. Turitsyn, E. G. Turitsyna, S. B. Medvedev, and M. P. Fedoruk, "Averaged model and integrable limits in nonlinear double-periodic Hamiltonian systems," *Physical Review E*, vol. 61, no. 3, pp. 3127–3132, March 2000.

[118] S. K. Turitsyn, M. P. Fedoruk, E. G. Shapiro, V. K. Mezenrsev, and E. G. Turitsyna, "Novel approaches to numerical modeling of periodic dispersion-managed fiber communication systems," *IEEE Journal of Selected Topics in Quantum Electronics*, vol. 6, no. 2, pp. 263–275, March/April 2000.

[119] J. P. Gordon and L. F. Mollenauer, "Phase noise in photonic communications systems using linear amplifiers," *Optics Letters*, vol. 15, no. 23, pp. 1351–1353, December 1990.

[120] C. Peucheret, N. Hanik, R. Freund, L. Molle, and P. Jeppesen, "Optimization of pre- and post-dispersion compensation schemes for 10-Gbits/s NRZ links using standard and dispersion compensating fibers," *IEEE Photonics Technology Letters*, vol. 12, no. 8, pp. 992–994, August 2000.

[121] M. Malach, "Faserunabhängiges Dispersionsmanagement zur Unterdrückung von SPM und XPM in 10 Gb/s NRZ-modulierten WDM-Übertragungssystemen," Ph.D. dissertation, Technische Universität Berlin, Berlin, Germany, November 2008.

[122] K. Inoue, "Phase-mismatching characteristic of four-wave mixing in fiber lines with multistage optical amplifiers," *Optics Letters*, vol. 17, no. 11, pp. 801–803, June 1992.

[123] E. A. Golovchenko, N. S. Bergano, and C. R. Davidson, "Four-wave mixing in multispan dispersion-managed transmission links," *IEEE Photonics Technology Letters*, vol. 10, no. 10, pp. 1481–1483, October 1998.

[124] J. K. Fischer, C.-A. Bunge, K. Jamshidi, H. Louchet, and K. Petermann, "Equivalent dispersion maps in fibre-optic communication systems," in *Proc. 32nd European Conference on Optical Communication, ECOC'06*, Cannes, France, September 2006, paper We3.P.134.

[125] J. K. Fischer, C.-A. Bunge, and K. Petermann, "Application of the nonlinear transfer function to dispersion map evaluation in DPSK transmission systems," in *Proc. 19th Annual Meeting of the IEEE Lasers and Electro-Optics Society, LEOS'06*, Montreal, Canada, November 2006, paper ThH2.

[126] J.-C. Antona, S. Bigo, and J.-P. Faure, "Nonlinear cumulated phase as a criterion to assess performance of terrestrial WDM systems," in *Proc. Optical Fiber Communication Conference, OFC'02*, Anaheim, USA, March 2002, paper WX5.

[127] S. Vorbeck and M. Schneiders, "Cumulative nonlinear phase shift as engineering rule for performance estimation in 160-Gb/s transmission systems," *IEEE Photonics Technology Letters*, vol. 16, no. 11, pp. 2571–2573, November 2004.

[128] B. Konrad and K. Petermann, "Optimum fiber dispersion in high-speed TDM systems," *IEEE Photonics Technology Letters*, vol. 13, no. 4, pp. 299–301, April 2001.

[129] A. Cauvin, Y. Frignac, and S. Bigo, "Nonlinear impairments at various bit rates in single-channel dispersion-managed systems," *Electronics Letters*, vol. 39, no. 23, pp. 1670–1671, November 2003.

[130] Y. Frignac, J.-C. Antona, and S. Bigo, "Enhanced analytical engineering rule for fast optimization of dispersion maps in 40 Gbit/s-based transmission," in *Proc. Optical Fiber Communication Conference, OFC'04*, Los Angeles, USA, February 2004, paper TuN3.

[131] B. Konrad, K. Petermann, J. Berger, R. Ludwig, C. Weinert, H. Weber, and B. Schmauss, "Impact of fiber chromatic dispersion in high-speed TDM transmission systems," *Journal of Lightwave Technology*, vol. 20, no. 12, pp. 2129–2135, December 2002.

[132] B. Konrad, "Dispersionsmanagement in optischen 160 Gbit/s Übertragungssystemen," Ph.D. dissertation, Technische Universität Berlin, Berlin, Germany, April 2004.

[133] Y. Frignac, J.-C. Antona, S. Bigo, and J.-P. Hamaide, "Numerical optimization of pre- and in-line dispersion compensation in dispersion-managed systems at 40 Gbit/s," in *Proc. Optical Fiber Communication Conference, OFC'02*, Anaheim, USA, March 2002, paper ThFF5.

[134] G. Charlet, "Progress in optical modulation formats for high-bit rate WDM transmissions," *IEEE Journal of Selected Topics in Quantum Electronics*, vol. 12, no. 4, pp. 469–483, July-August 2006.

[135] P. J. Winzer and R.-J. Essiambre, "Advanced Modulation Formats for High-Capacity Optical Transport Networks," *Journal of Lightwave Technology*, vol. 24, no. 12, pp. 4711–4728, December 2006.

[136] S. Savory and A. Hadjifotiou, "Laser linewidth requirements for optical DQPSK systems," *IEEE Photonics Technology Letters*, vol. 16, no. 3, pp. 930–932, March 2004.

[137] N. S. Avlonitis and E. M. Yeatman, "Performance evaluation of optical DQPSK using saddle point approximation," *Journal of Lightwave Technology*, vol. 24, no. 3, pp. 1176–1185, March 2006.

[138] G. L. Li and P. K. L. Yu, "Optical intensity modulators for digital and analog applications," *Journal of Lightwave Technology*, vol. 21, no. 9, pp. 2010–2030, September 2003.

[139] D. Fonseca, A. V. T. Cartaxo, and P. P. Monteiro, "Influence of the extinction ratio on the intrachannel nonlinear distortion of 40-Gb/s return-to-zero transmission systems over standard fiber," *Journal of Lightwave Technology*, vol. 25, no. 6, pp. 1447–1457, June 2007.

[140] ITU-T, "G.959.1 Optical transport network physical layer interfaces."

[141] R. A. Griffin and A. C. Carter, "Optical differential quadrature phase-shift key (oDQPSK) for high capacity optical transmission," in *Proc. Optical Fiber Communication Conference, OFC'02*, Anaheim, USA, March 2002, paper WX6.

[142] W. Kaiser, T. Wuth, M. Wichers, and W. Rosenkranz, "Reduced complexity optical duobinary 10-Gb/s transmitter setup resulting in an increased transmission distance," *IEEE Photonics Technology Letters*, vol. 13, no. 8, pp. 884–886, August 2001.

[143] C. Rasmussen, T. Fjelde, J. Bennike, F. Liu, S. Dey, B. Mikkelsen, P. Mamyshev, P. Serbe, P. van der Wagt, Y. Akasaka, D. Harris, D. Gapontsev, V. Ivshin, and P. Reeves-Hall, "DWDM 40 G transmission over trans-Pacific distance (10,000 km) using CSRZ-DPSK, enhanced FEC and all-Raman amplified 100 km UltraWave fiber spans," in *Proc. Optical Fiber Communications Conference, OFC'03*, Atlanta, USA, March 2003, paper PD18.

[144] C. Wree, J. Leibrich, and W. Rosenkranz, "RZ-DQPSK Format with High Spectral Efficiency and High Robustness Towards Fiber Nonlinearities," in *Proc. 28th European Conference on Optical Communication, ECOC'02*, Copenhagen, Denmark, September 2002, paper 9.6.6.

[145] N. G. de Bruijn, "A combinatorial problem," *Proc. Koninklijke Nederlandse Akademie van Wetenschappen*, vol. 49, pp. 758–764, 1946.

[146] P. Serena, A. Orlandini, and A. Bononi, "The memory of optimized dispersion-managed periodic optical links," in *Proc. 33rd European Conference on Optical Communication, ECOC'07*, Berlin, Germany, September 2007, paper P093.

[147] J.-C. Antona, E. Grellier, A. Bononi, S. Petitrenaud, and S. Bigo, "Revisiting binary sequence length requirements for the accurate emulation of highly

dispersive transmission systems," in *Proc. 34th European Conference on Optical Communication, ECOC'08*, Brussels, Belgium, September 2008, paper We.1.E.3.

[148] B. Spinnler and C. Xie, "Performance assessment of DQPSK using pseudo-random quaternary sequences," in *Proc. 33rd European Conference on Optical Communication, ECOC'07*, Berlin, Germany, September 2007, paper 2.3.6.

[149] D. van den Borne, E. Gottwald, G. D. Khoe, and H. D. Waardt, "Bit pattern dependence in optical DQPSK modulation," *Electronics Letters*, vol. 43, no. 22, pp. 1223–1225, October 2007.

[150] E. Yamazaki, F. Inuzuka, K. Yonenaga, A. Takada, and M. Koga, "Compensation of interchannel crosstalk induced by optical fiber nonlinearity in carrier phase-locked WDM system," *IEEE Photonics Technology Letters*, vol. 19, no. 1, pp. 9–11, January 2007.

[151] P. J. Winzer, M. Pfennigbauer, and R.-J. Essiambre, "Coherent crosstalk in ultradense WDM systems," *Journal of Lightwave Technology*, vol. 23, no. 4, pp. 1734–1744, April 2005.

[152] M. G. Taylor, "Coherent detection method using DSP for demodulation of signal and subsequent equalization of propagation impairments," *IEEE Photonics Technology Letters*, vol. 16, no. 2, pp. 674–676, February 2004.

[153] A. Leven, N. Kaneda, U.-V. Koc, and Y.-K. Chen, "Coherent receivers for practical optical communication systems," in *Proc. Optical Fiber Communication Conference and National Fiber Optic Engineers Conference*, Anaheim, USA, March 2007, paper OThK4.

[154] S. Randel, "Analyse faseroptischer Übertragungssysteme mit Wellenlängenmultiplex bei 160 Gbit/s Kanaldatenrate," Ph.D. dissertation, Technische Universität Berlin, Berlin, Germany, May 2005.

[155] P. J. Winzer, M. Pfennigbauer, M. M. Strasser, and W. R. Leeb, "Optimum filter bandwidths for optically preamplified NRZ receivers," *Journal of Lightwave Technology*, vol. 19, no. 9, pp. 1263–1273, September 2001.

[156] P. J. Winzer, S. Chandrasekhar, and H. Kim, "Impact of filtering on RZ-DPSK reception," *IEEE Photonics Technology Letters*, vol. 15, no. 6, pp. 840–842, June 2003.

[157] P. J. Winzer and H. Kim, "Degradations in balanced DPSK receivers," *IEEE Photonics Technology Letters*, vol. 15, no. 9, pp. 1282–1284, September 2003.

[158] G. Bosco and P. Poggiolini, "On the joint effect of receiver impairments on direct-detection DQPSK systems," *Journal of Lightwave Technology*, vol. 24, no. 3, pp. 1323–1333, March 2006.

[159] A. H. Gnauck and P. J. Winzer, "Optical Phase-Shift-Keyed Transmission," *Journal of Lightwave Technology*, vol. 23, no. 1, pp. 115–130, January 2005.

[160] E. Forestieri, "Evaluating the error probability in lightwave systems with chromatic dispersion, arbitrary pulse shape and pre- and postdetection filtering," *Journal of Lightwave Technology*, vol. 18, no. 11, pp. 1493–1503, November 2000.

[161] J. Wang and J. M. Kahn, "Impact of chromatic and polarization-mode dispersions on DPSK systems using interferometric demodulation and direct detection," *Journal of Lightwave Technology*, vol. 22, no. 2, pp. 362–371, February 2004.

[162] C. R. Davidson, C. J. Chen, M. Nissov, A. Pilipetskii, N. Ramanujam, H. D. Kidorf, B. Pedersen, M. A. Mills, C. Lin, M. I. Hayee, J. X. Cai, A. B. Puc, P. C. Corbett, R. Menges, H. Li, A. Elyamani, C. Rivers, and N. S. Bergano, "1800 Gb/s transmission of one hundred and eighty 10 Gb/s WDM channels over 7000 km using the full EDFA C-band," in *Proc. Optical Fiber Communication Conference, OFC'00*, Baltimore, USA, March 2000, paper PD25-1.

[163] J.-X. Cai, M. Nissov, C. R. Davidson, Y. Cai, A. N. Pilipetskii, H. Li, M. A. Mills, R.-M. Mu, U. Feiste, L. Xu, A. J. Lucero, D. Foursa, and N. Bergano, "Transmission of thirty-eight 40 Gb/s channels (>1.5 Tb/s) over transoceanic distance," in *Proc. Optical Fiber Communication Conference, OFC'02*, Anaheim, USA, March 2002, paper FC4-1.

[164] E. Desurvire, "Spectral noise figure of Er^{3+}-doped fiber amplifiers," *IEEE Photonics Technology Letters*, vol. 2, no. 3, pp. 208–210, March 1990.

[165] C.-A. Bunge, J. K. Fischer, H. Louchet, and K. Petermann, "The nonlinear diffusion bandwidth for characterizing optical fibre transmission systems," in *Proc. 8th International Conference on Transparent Optical Networks, ICTON'06*, Nottingham, United Kingdom, June 2006, paper Mo.C1.5.

[166] J. K. Fischer, C.-A. Bunge, K. Jamshidi, H. Louchet, and K. Petermann, "Nonlinear diffusion bandwidth and equivalent single-span model: Simple tools for the optimization of dispersion maps," in *Proc. 11th European Conference on Networks and Optical Communications, NOC'06*, Berlin, Germany, July 2006, pp. 97–104.

[167] C.-A. Bunge, J. K. Fischer, H. Louchet, K. Jamshidi, and K. Petermann, "The nonlinear diffusion bandwidth: A simple tool for optimizing dispersion maps," in *Proc. Asia-Pacific Optical Communications, APOC'06*, Gwangju, South Korea, September 2006, paper 6353-67.

[168] E. Ip and J. M. Kahn, "Power Spectra of Return-to-Zero Optical Signals," *Journal of Lightwave Technology*, vol. 24, no. 3, pp. 1610–1618, March 2006.

[169] A. Bononi, P. Serena, and A. Orlandini, "A unified design framework for single-channel dispersion-managed terrestrial systems," *Journal of Lightwave Technology*, vol. 26, no. 22, pp. 3617–3631, November 2008.

[170] C. Xie, L. F. Mollenauer, and N. Mamysheva, "Numerical study of random variations of span lengths and span path-average dispersions on dispersion-managed soliton system performance," *Journal of Lightwave Technology*, vol. 21, no. 3, pp. 769–775, March 2003.

[171] D. C. Kilper, S. Chandrasekhar, E. Burrows, L. L. Buhl, and J. Centanni, "Local dispersion map deviations in metro-regional transmission investigated using a dynamically re-configurable re-circulating loop," in *Proc. Optical Fiber Communications Conference and National Fiber Optic Engineers Conference, OFC/NFOEC'07*, Anaheim, USA, March 2007, paper OThL5.

[172] P. A. Humblet and M. Azizoglu, "On the bit error rate of lightwave systems with optical amplifiers," *Journal of Lightwave Technology*, vol. 9, no. 11, pp. 1576–1582, November 1991.

[173] X. Liu, C. Xu, and X. Wei, "Nonlinear phase noise in pulse-overlapped transmission based on return-to-zero differential-phase-shift-keying," in *Proc. 28th*

European Conference on Optical Communication, ECOC'02, Copenhagen, Denmark, September 2002, paper 9.6.5.

[174] M. Rohde, C. Caspar, N. Heimes, M. Konitzer, E.-J. Bachus, and N. Hanik, "Robustness of DPSK direct detection transmission format in standard fibre WDM systems," *Electronics Letters*, vol. 36, no. 17, pp. 1483–1484, August 2000.

[175] J. Leibrich, C. Wree, and W. Rosenkranz, "CF-RZ-DPSK for suppression of XPM on dispersion-managed long-haul optical WDM transmission on standard single-mode fiber," *IEEE Photonics Technology Letters*, vol. 14, no. 2, pp. 155–157, February 2002.

[176] C. Xu, X. Liu, L. Mollenauer, and X. Wei, "Comparison of return-to-zero differential phase-shift keying and ON-OFF keying in long-haul dispersion managed transmission," *IEEE Photonics Technology Letters*, vol. 15, no. 4, pp. 617–619, April 2003.

[177] X. Wei and X. Liu, "Analysis of intrachannel four-wave mixing in differential phase-shift keying transmission with large dispersion," *Optics Letters*, vol. 28, no. 23, pp. 2300–2302, December 2003.

[178] R. Dischler, A. Klekamp, J. Lazaro, and W. Idler, "Experimental comparison of non linear threshold and optimum pre dispersion of 43 Gb/s ASK and DPSK formats," in *Proc. Optical Fiber Communication Conference, OFC'04*, Los Angeles, USA, February 2004, paper TuF4.

[179] A. Klekamp, R. Dischler, and W. Idler, "Fibre nonlinear threshold comparison of SMF and NZDSF types on binary 43 Gb/s modulation formats," in *Proc. 31st European Conference on Optical Communication, ECOC'05*, Glasgow, Scotland, September 2005, paper We4.P.059.

[180] ——, "DWDM and single channel fibre nonlinear thresholds for 43 Gb/s ASK and DPSK formats over various fibre types," in *Proc. Optical Fiber Communication Conference and National Fiber Optic Engineers Conference, OFC/NFOEC'06*, Anaheim, USA, March 2006, paper OFD5.

[181] J. K. Fischer, S. Randel, A. Denzin, and K. Petermann, "Optimum fibre dispersion for 40 Gbit/s DPSK transmission," in *Proc. 31st European Conference*

on *Optical Communication, ECOC'05*, Glasgow, Scotland, September 2005, paper Tu1.2.5.

[182] E. Pincemin, A. Tan, A. Tonello, S. Wabnitz, J. D. Ania-Castañón, V. Mezentsev, S. K. Turitsyn, Y. Jaouën, and L. Grüner-Nielsen, "Performance comparison of SSMF and UltraWave fibers for ultra-long-haul 40-Gb/s WDM transmission," *IEEE Photonics Technology Letters*, vol. 19, no. 20, pp. 1613–1615, October 2007.

[183] H. Kim and R.-J. Essiambre, "Transmission of 8×20 Gb/s DQPSK signals over 310-km SMF with 0.8-b/s/Hz spectral efficiency," *IEEE Photonics Technology Letters*, vol. 15, no. 5, pp. 769–771, May 2003.

[184] N. Yoshikane and I. Morita, "1.14 b/s/Hz spectrally efficient 50×85.4-Gb/s transmission over 300 km using copolarized RZ-DQPSK signals," *Journal of Lightwave Technology*, vol. 23, no. 1, pp. 108–114, January 2005.

[185] G. Charlet, P. Tran, H. Mardoyan, M. Lefrancois, T. Fauconnier, F. Jorge, and S. Bigo, "151×43 Gb/s transmission over 4080 km based on return-to-zero-differential quadrature phase-shift keying," in *Proc. 31st European Conference on Optical Communication, ECOC'05*, Glasgow, Scotland, September 2005, paper Th4.1.3.

[186] M. Daikoku, I. Morita, H. Taga, H. Tanaka, T. Kawanishi, T. Sakamoto, T. Miyazaki, and T. Fujita, "100-Gb/s DQPSK transmission experiment without OTDM for 100G Ethernet transport," *Journal of Lightwave Technology*, vol. 25, no. 1, pp. 139–145, January 2007.

[187] A. H. Gnauck, P. J. Winzer, C. Dorrer, and S. Chandrasekhar, "Linear and nonlinear performance of 42.7-Gb/s single-polarization RZ-DQPSK format," *IEEE Photonics Technology Letters*, vol. 18, no. 7, pp. 883–885, April 2006.

[188] O. Vassilieva, T. Hoshida, S. Choudhary, and H. Kuwahara, "Nonlinear tolerant and spectrally efficient 86 Gbit/s RZ-DQPSK format for a system upgrade," in *Proc. Optical Fiber Communication Conference, OFC'03*, Georgia, USA, March 2003, paper ThE7.

[189] H. Kim and A. Gnauck, "Experimental investigation of the performance limitation of DPSK systems due to nonlinear phase noise," *IEEE Photonics Technology Letters*, vol. 15, no. 2, pp. 320–322, February 2003.

[190] T. Mizuochi, K. Ishida, T. Kobayashi, J. Abe, K. Kinjo, K. Motoshima, and K. Kasahara, "A comparative study of DPSK and OOK WDM transmission over transoceanic distances and their performance degradations due to nonlinear phase noise," *Journal of Lightwave Technology*, vol. 21, no. 9, pp. 1933–1943, September 2003.

[191] C. Xu, X. Liu, and X. Wei, "Differential phase-shift keying for high spectral efficiency optical transmissions," *IEEE Journal of Selected Topics in Quantum Electronics*, vol. 10, no. 2, pp. 281–293, March-April 2004.

[192] M. Ohm, R.-J. Essiambre, and P. J. Winzer, "Nonlinear phase noise and distortion in 42.7-Gbit/s RZ-DPSK systems," in *Proc. 31st European Conference on Optical Communication, ECOC'05*, Glasgow, Scotland, September 2005, paper Tu1.2.1.

[193] P. Serena, A. Orlandini, and A. Bononi, "Parametric-gain approach to the analysis of single-channel DPSK/DQPSK systems with nonlinear phase noise," *Journal of Lightwave Technology*, vol. 24, no. 5, pp. 2026–2037, May 2006.

[194] A. Demir, "Nonlinear Phase Noise in Optical-Fiber-Communication Systems," *Journal of Lightwave Technology*, vol. 25, no. 8, pp. 2002–2032, August 2007.

[195] M. Seimetz, *High-order modulation for optical fiber transmission*, ser. Springer Series in Optical Sciences. Berlin / Heidelberg: Springer, July 2009, vol. 143.

[196] J. McNicol, M. O'Sullivan, K. Roberts, A. Comeau, D. McGhan, and L. Strawczynski, "Electrical domain compensation of optical dispersion," in *Proc. Optical Fiber Communication Conference and National Fiber Optic Engineers Conference, OFC/NFOEC'05*, Anaheim, USA, March 2005, paper OThJ3.

[197] D. McGhan, C. Laperle, A. Savchenko, C. Li, G. Mak, and M. O'Sullivan, "5120 km RZ-DPSK transmission over G652 fiber at 10 Gb/s with no optical dispersion compensation," in *Proc. Optical Fiber Communication Conference and National Fiber Optic Engineers Conference, OFC/NFOEC'05*, Anaheim, USA, March 2005, paper PDP27.

[198] R. I. Killey, P. M. Watts, M. Glick, and P. Bayvel, "Electronic precompensation techniques to combat dispersion and nonlinearities in optical transmission," in *Proc. 31st European Conference on Optical Communication, ECOC'05*, Glasgow, Scotland, September 2005, paper Tu4.2.1.

[199] M. S. O'Sullivan, K. Roberts, and C. Bontu, "Electronic dispersion compensation techniques for optical communication systems," in *Proc. 31st European Conference on Optical Communication, ECOC'05*, Glasgow, Scotland, September 2005, paper Tu3.2.1.

[200] D. McGhan, C. Laperle, A. Savchenko, C. Li, G. Mak, and M. O'Sullivan, "5120-km RZ-DPSK transmission over G.652 fiber at 10 Gb/s without optical dispersion compensation," *IEEE Photonics Technology Letters*, vol. 18, no. 2, pp. 400–402, January 2006.

[201] K. Roberts, C. Li, L. Strawczynski, M. O'Sullivan, and I. Hardcastle, "Electronic precompensation of optical nonlinearity," *IEEE Photonics Technology Letters*, vol. 18, no. 2, pp. 403–405, January 2006.

[202] C. Weber, J. K. Fischer, C.-A. Bunge, and K. Petermann, "Electronic precompensation of intrachannel nonlinearities at 40 Gb/s," *IEEE Photonics Technology Letters*, vol. 18, no. 16, pp. 1759–1761, August 2006.

[203] C. Xie and R.-J. Essiambre, "Intra-channel nonlinearity compensation in 40-Gb/s systems by combining electronic pre-distortion and optical dispersion compensation," in *Proc. 32nd European Conference on Optical Communication, ECOC'06*, Cannes, France, September 2006, paper We3.P.82.

[204] C. Xie, "Performance of electronic pre-distortion in 40-Gb/s systems with optical dispersion compensation for different modulation formats and transmission fibres," in *Proc. 33rd European Conference on Optical Communication, ECOC'07*, Berlin, Germany, September 2007, paper 3.1.5.

[205] R.-J. Essiambre, G. J. Foschini, P. J. Winzer, G. Kramer, and E. C. Burrows, "The capacity of fiber-optic communication systems," in *Proc. Optical Fiber Communication Conference and National Fiber Optic Engineers Conference, OFC/NFOEC'08*, San Diego, USA, February 2008, paper OTuE1.

[206] C. Paré, A. Villeneuve, P.-A. Bélanger, and N. J. Doran, "Compensating for dispersion and the nonlinear Kerr effect without phase conjugation," *Optics Letters*, vol. 21, no. 7, pp. 459–461, April 1996.

[207] R.-J. Essiambre, P. J. Winzer, X. Q. Wang, W. Lee, C. A. White, and E. C. Burrows, "Electronic Predistortion and Fiber Nonlinearity," *IEEE Photonics Technology Letters*, vol. 18, no. 17, pp. 1041–1135, September 2006.

[208] J. F. Elder, "Global R^d optimization when probes are expensive: the GROPE algorithm," in *Proc. IEEE International Conference on Systems, Man and Cybernetics*, vol. 1, Chicago, USA, October 1992, pp. 577–582.

[209] O. Gaete, L. D. Coelho, B. Spinnler, E.-D. Schmidt, and N. Hanik, "Global optimization of optical communication systems," in *Proc. 34th European Conference on Optical Communication, ECOC'08*, Brussels, Belgium, September 2008, paper P.4.17.

[210] X. Li, X. Chen, G. Goldfarb, E. Mateo, I. Kim, F. Yaman, and G. Li, "Electronic post-compensation of WDM transmission impairments using coherent detection and digital signal processing," *Optics Express*, vol. 16, no. 2, pp. 880–888, January 2008.

[211] G. Goldfarb and G. Li, "Demonstration of fibre impairment compensation using split-step infinite-impulse-response filtering method," *Electronics Letters*, vol. 44, no. 13, pp. 814–816, June 2008.

[212] E. Ip and J. M. Kahn, "Compensation of dispersion and nonlinear impairments using digital backpropagation," *Journal of Lightwave Technology*, vol. 26, no. 20, pp. 3416–3425, October 2008.

[213] K. Kikuchi, "Phase-diversity homodyne detection of multilevel optical modulation with digital carrier phase estimation," *IEEE Journal of Selected Topics in Quantum Electronics*, vol. 12, no. 4, pp. 563–570, July/August 2006.

[214] C. R. Menyuk and B. S. Marks, "Interaction of Polarization Mode Dispersion and Nonlinearity in Optical Fiber Transmission Systems," *Journal of Lightwave Technology*, vol. 24, no. 7, pp. 2806–2826, July 2006.

Die VDM Verlagsservicegesellschaft sucht für wissenschaftliche Verlage abgeschlossene und herausragende

Dissertationen, Habilitationen, Diplomarbeiten, Master Theses, Magisterarbeiten usw.

für die kostenlose Publikation als Fachbuch.

Sie verfügen über eine Arbeit, die hohen inhaltlichen und formalen Ansprüchen genügt, und haben Interesse an einer honorarvergüteten Publikation?

Dann senden Sie bitte erste Informationen über sich und Ihre Arbeit per Email an *info@vdm-vsg.de*.

Sie erhalten kurzfristig unser Feedback!

VDM Verlagsservicegesellschaft mbH
Dudweiler Landstr. 99
D - 66123 Saarbrücken

Telefon +49 681 3720 174
Fax +49 681 3720 1749

www.vdm-vsg.de

Die VDM Verlagsservicegesellschaft mbH vertritt

Printed by Books on Demand GmbH, Norderstedt / Germany